Core Books in Advanced Mathematics

Examinations in Mathematics

Core Books in Advanced Mathematics

General Editor: C. PLUMPTON, Moderator in Mathematics,
University of London School Examinations Board;
formerly Reader in Engineering Mathematics,
Queen Mary College, University of London.

Titles available:

Differentiation
Integration
Vectors
Curve Sketching
Newton's Laws and Particle Motion
Mechanics of Groups of Particles
Methods of Trigonometry
Coordinate Geometry and Complex Numbers
Proof
Statistics
Probability
Methods of Algebra
Examinations in Mathematics

Core Books in Advanced Mathematics

Examinations in Mathematics

Eva Shipton
Chief Examiner in Mathematics,
University of London School Examinations Board;
formerly Deputy Head, Owen's School, Potters Bar.

Michael Kenwood
Director of Studies, King Edward's School, Bath;
Chief Examiner in Advanced-level Mathematics,
University of London School Examinations Board.

Cyril Moss
Formerly Senior Lecturer and Deputy Head of Mathematics,
The City University, London;
Chief Examiner in Advanced-level Mathematics,
University of London School Examinations Board.

Charles Plumpton
Moderator in Mathematics, University of
London School Examinations Board;
formerly Reader in Engineering Mathematics,
Queen Mary College, University of London.

Macmillan Education
London and Basingstoke

First published 1985

Published by
MACMILLAN EDUCATION LTD
Houndmills, Basingstoke, Hampshire RG21 2XS
and London
Companies and representatives
throughout the world

Typeset by Mid-County Press, London SW15
Printed in Hong Kong

British Library Cataloguing in Publication Data

Examinations in mathematics.————(Core books in
advanced mathematics)
1. Mathematics————Study and teaching (Secondary)
————Great Britain   2. Mathematics————Examinations
I. Shipton, E.   II. Series
510'.76   QA14.G7
ISBN 0-333-39455-0

Contents

Preface

Advanced level mathematics syllabuses are once again undergoing changes in content and approach following the revolution in the early 1960s which led to the unfortunate dichotomy between 'modern' and 'traditional' mathematics. The current trend in syllabuses for Advanced level mathematics now being developed and published by many GCE Boards is towards an integrated approach, taking the best of the topics and approaches of modern and traditional mathematics, in an attempt to create a realistic examination target through syllabuses which are maximal for examining and minimal for teaching. In addition, resulting from a number of initiatives, core syllabuses are being developed for Advanced level mathematics consisting of techniques of pure mathematics as taught in schools and colleges at this level.

The concept of a core can be used in several ways, one of which is mentioned above, namely the idea of a core syllabus to which options such as theoretical mechanics, further pure mathematics and statistics can be added. The books in this series involve a different use of the core idea. They are books on a range of topics, each of which is central to the study of Advanced level mathematics; they form small studies of their own, of topics which together cover the main areas of any single-subject mathematics syllabus at Advanced level.

Particularly at times when economic conditions make the problems of acquiring comprehensive textbooks giving complete syllabus coverage acute, schools and colleges and individual students can collect as many of the core books as they need to supplement books they already have, so that the most recent syllabuses of, for example, the London, Cambridge, AEB and JMB GCE Boards can be covered at minimum expense. Alternatively, of course, the whole set of core books gives complete syllabus coverage of single-subject Advanced level mathematics syllabuses.

The aim of each book is to develop a major topic of the single-subject syllabuses, giving essential book work, worked examples and numerous exercises arising from the authors' vast experience of examining at this level. Thus, as well as using the core books in either of the above ways, they are ideal for supplementing comprehensive textbooks, by providing more examples and exercises, so necessary for the preparation and revision for examinations.

The subject of this volume of the series is examinations in mathematics. Even in the later stages of a mathematical course, few pupils and not all teachers are aware of the marking techniques used by examiners in public examinations. A carefully planned strategy and an awareness of the basic ideas of mathematical thinking, coupled with understanding of marking processes, can vastly improve the performance of candidates in examinations.

The present book has been written in an attempt to assist sixth form teachers and pupils to maximize their potential in pure mathematics and applied mathematics, including probability and statistics. Experience suggests that students at colleges of further and higher education could also benefit from a study of this book.

The various types of tests used by examination boards are discussed in detail and solutions and marking schemes are given and explained in great detail. Additional papers are given for students to try and, it is hoped, discuss with one another.

Finally, a chapter on revision methods is intended to guide students along correct lines for both school-based and external examinations.

Eva Shipton
Michael Kenwood
Cyril Moss
Charles Plumpton

1 Examination strategy

1.1 General points

You should have seen some old (past) papers set on the particular syllabus for which you have entered, or, even if you are taking a new syllabus, the examination board will have published a specimen paper for your information. Study the papers carefully and take particular note of the *rubric*, that is the instructions at the start of the paper telling you how many questions to attempt, how much time you are allowed to answer the paper, what materials (formulae book, tables etc.) are supplied, and whether you should start the answer to each question on a fresh page. For example, if the rubric states 'In calculations you are advised to show intermediate steps and answers in your working,' it means precisely *that*, and if you do not show intermediate steps then marks are very likely to be lost; and, if just a final *wrong* answer is shown (everything being done by calculator), then you will gain no marks at all.

Obey the instructions. Remember that examiners have to read several hundred scripts and that, while they try to treat every candidate fairly and equally, they are bound to be influenced by a clearly written script in which the rules have been obeyed. Remember also that, unless you are instructed to the contrary, you may answer the questions in any order.

In some cases you can check your answer(s) very simply. For example, if you have to (i) find an integral or (ii) resolve an algebraic expression into partial fractions, a simple check by differentiating your answer to (i) or recombining your results for (ii) takes little time and effort and gives you reassurance. Answers to inequalities can sometimes be partially checked by using trial values. Equations relating physical quantities should be dimensionally correct, i.e. each (complete) term in an equation must have the same dimension as the others. For example, the equations $Pt = mv$ and $Ps = \frac{1}{2}mv^2$, giving the velocity v of a body of mass m acted upon by a constant force P for a time t while covering a distance s, are both dimensionally correct, but the equation $Pt = \frac{1}{2}mv^2$ cannot be correct. Equations derived by taking moments can sometimes be partially checked by using trial values for the angles involved. However, judgment must be exercised on whether time can be spared for checking. It should not be done at the expense of not completing the paper.

1.2 Using the time allowed to the best advantage

Do not waste time copying out the question. (The examiner has a copy of the paper!) All you need do is to put down the number of the question on your script where required.

Do not waste time underlining or writing 'Q.E.D.'

Do, where appropriate, give fairly large, clearly labelled freehand diagrams. Accurate pencil figures constructed using compasses, flexi-curves and the like for circles, parabolas etc. are seldom needed. *Do* learn to sketch circles, straight lines and other well-known curves quickly and reasonably accurately without using drawing instruments. The use of drawing instruments, except perhaps a ruler, is not necessary. *Do* write reasonably clearly, space out your work and, provided your writing is fairly clear, that should be sufficient. No credit is given for copperplate writing.

Do not waste time at the end of the examination writing an essay informing the examiner how you would complete the questions if you had time. Such essays do not earn you any marks, for the examiner cannot give credit for efforts to bluff him/her that you could answer more questions or complete some part answers if more time were available. The examiner knows how to do the questions and is only interested in *your* solutions. Even if you only have five minutes or so left, continue writing down what you think are the solutions and so continue the possibility of earning marks until time is up. Do not waste time on any confidence tricks—you will receive marks only for the relevant mathematics which you write down and for nothing else. Irrelevant appeals for mercy, goodwill etc. fall on deaf ears.

1.3 Answering the paper

Glance through the question paper before starting to write. You will then get a feel for the demands of each question, and should gain an impression of those questions you feel that you should attempt while eliminating those questions which you feel are 'not for you', particularly those with later parts you do not understand or will be unable to complete. Of course, in mathematics you may not be sure that you can do a question until you have had a go. In particular, answer the questions set and *not* those you would have liked. You receive no credit for setting and answering your own paper! Experience shows that more marks are earned by a few complete questions than by many fragments.

Where you have a choice (say in a particular section of the paper you have to answer four questions from six) answer four only. Do not dissipate your effort by answering five or six. However, if you are dissatisfied with one of the questions you have attempted and have time, you might find it worth while to try another one. Some examination boards give credit for the four best answers, in which case you would leave both your attempts. Otherwise, cross out the one you think the weaker. *Do not* cross out an attempt until you have replaced it with what you consider is a better attempt.

Do not fake a given answer, that is fraudulently state that you have proved a result when you really have not done so. When an answer is given, the examiner makes certain that a solution is not faked. If an answer is faked and there is doubt regarding the mark for a subsequent question, the examiner will tend *not* to give the candidate the benefit of the doubt, as he/she might otherwise do. For example, examiners dislike seeing symbols such as $\pm$ which could be positive or negative, and suggests that the candidate does not know which is correct.

A vital point is 'knowing when to stop' in the working of a question when the manipulation gets more and more difficult and involved. It is safe to say that examiners, nowadays, do not set questions involving very lengthy manipulation. If you get the feeling of being 'bogged down' in heavy algebra, *stop*. Look back to the beginning of the question to make sure that you are tackling it in the right way and check through the early working to see whether you have made an error. If you can find nothing wrong, leave the question and try another one but do not cross out the work you have done as it may be worth some method marks and you may wish to have another look at it if you have time later.

If, after checking your work, you are still unable to find an error in an attempt to obtain a given answer, then it is often advisable to start working the question again from the beginning. Errors which you have passed over on checking usually come to light this way.

When marks or part-marks for questions are shown, these give a rough indication of the length of time you should take to answer them. Do not spend too much time for few marks. For example, four marks (out of 100 in a paper of two and a half hours' duration) suggests a time of about six minutes to answer. Spending more than ten minutes in such a case is a waste and suggests that either you have the wrong method or you have made mistakes. Leave the question and try another—this way you will not panic.

Long questions should not be feared. These are usually structured and so the examiner is leading you through the process which you should follow. Perhaps there is an intermediate answer to be proved, that is an intermediate result is printed on the paper. If you do not or cannot get this intermediate result correct, do not continue with *your* result but assume that the printed result is, in fact, correct (making this clear on your script) and continue with this to do the rest of the question. This will almost certainly lead to easier working and, moreoever, ensure that you can also get full marks for the work you do correctly in the remainder of the question, provided that you accept the examiner's answer as correct and continue from there. If you *misread* (MR) a question, without making it appreciably easier, and you perform similar stages to those intended, it is still possible to receive most of the marks for the following parts which you do correctly (following your misread). But try to avoid misreading—it usually makes the question harder.

Sometimes a question says 'Prove by . . . , *or otherwise,* . . . .' This means that the examiner is suggesting a firm method of attack but that you may use some other method if you see or know one. However, the 'or otherwise' is rarely suitable. Frequently, it means that the examiner is being cautious in case there is a better method which he does not know!

When there is a choice of questions, make sure you answer first the questions you think you can do best. *Do not* make the mistake of saving your best questions for the last part of the time allowed. If your preparation has been reasonable and conscientious, you will find that you recognize many of the questions. Rarely can examiners set 'new' questions—they usually produce rehashed 'old' questions with small 'twists' and different numerical values. Look to see which of the questions you have seen before and which you know how to do. Try this with a few old papers. You will be surprised to notice how many questions in your examination paper you have already seen, perhaps thinly disguised, in your textbooks.

However, in the type of paper where there is no choice and all the questions must be attempted, it is much better to work straight through the paper. There are two reasons why this is better: (i) it is time-saving, (ii) such papers have been carefully graded by examiners with the easier questions first, which means that candidates should have built up confidence by the time they reach the harder questions at the end.

Remember to use all the given information in any question. Rarely do examiners give superfluous (unnecessary) information and it is most unlikely that you will be able to complete a question without using all the given data.

1.4 Calculators

You must be familiar and expert with your calculator. It should contain the mathematical functions and operations occurring most frequently. Confidence and accuracy in its use are invaluable, indeed are essential.

When you use a calculator, write 'By calculator' at the appropriate places, and, if instructed, give intermediate expressions and/or answers where appropriate. Remember that a calculator may give numerical results to eight or nine significant figures and you may have to correct to three or four significant figures. There is no sense in giving an answer to eight significant figures when the data in the question is only accurate to three or four significant figures.

Always check at the end of any solution in which a numerical answer is required that you have answered the question set to the degree of accuracy specified. If a question states that an answer is required to two decimal places and you ignore this request, then at least one mark will be lost, even though you may have performed everything correctly through the solution (except, of course, doing as you are told at the end). Sometimes, the degree of accuracy required is only written as a general request in the rubric at the front of the examination paper, for example, 'Give answers to three significant figures unless told otherwise in a particular question.' Sometimes numerical approximations of constants, such as π, or the magnitude of g, the acceleration due to gravity, or the value of e, the base of natural logarithms, are given in this way and it should be part of your preparation to note the policy being adopted by your examiners.

1.5 Repeated attempts

If you try a question several times, do give some indication that the question is continued later. You may find that you do the first part of a question, then fail to complete the next part. But later perhaps you see what to do, or have time to try it again. Give some indication such as 'continued on page xx or later'. Better still, leave a page or part of a page to complete the unfinished question. Indicate, however, that your next question is on 'page xx'.

1.6 Preparation just before the examination

Revision is a very individual matter about which it is difficult to lay down general rules. However, in mathematics there is really no substitute for steady work throughout the course. A famous mathematician once said, 'The royal road to mathematics is to do ten examples every day and eleven on Sundays!' Frantic work in the last week or two will not make up for lack of steady application over the previous two years. Last-minute revision is of doubtful value at best, although some people like to brush up important points and formulae on the day before an examination.

To perform well in a mathematics paper you must be fresh and not worn out by last minute swotting. Not much 'memory work' is required in Advanced level mathematics. As suggested earlier, you need to be able to recognize those questions you have seen before and then apply the processes of mathematical thinking outlined in Chapter 2.

Do not spend the evening, before a mathematics paper, studying. It will *not* help you to tire yourself with last minute efforts at revision. Rather, have a relaxing evening and a good night's sleep. Enter the examination room relaxed and confident.

Experience shows that candidates tend to underperform on the first paper of a public examination. They become excited, panic if things do not run very smoothly, and make silly mistakes. A good way of avoiding such a calamity is to obtain an unseen paper on your syllabus, and try it out under examination conditions, in the allotted time, a few days (one or two!) before the examination proper. Work yourself up over this paper, get rid of your stresses and panics, and reflect on the mistakes you have made. Then in the operational examination you can keep cool and collected and maximize your potential.

1.7 Examination format

In Chapters 4 to 7 we illustrate three approaches in current use by which your knowledge of mathematics may be tested.

(i) *Objective tests (multiple-choice tests)*

In an objective test, you are required to select the correct answer, or the correct combination of answers, from a list of possibilities which follow the question. At present, not all examination boards use these; check your syllabus carefully. If you are required to take this form of examination, *read the rubric with great care*. Chapters 4 and 5 give explanations of the techniques required for answering multiple-choice tests. Papers 1 and 3 are typical multiple-choice tests at this level and an explanation of the solutions and mode of thinking required is given. Papers 2 and 4 are included for you to attempt on your own.

(ii) *Short questions and questions of variable length (Chapters 6 and 7)*

Each short question is designed to test one or two principles. In most papers using this approach, you are required to answer ALL of the questions set. This approach gives the examiner a way of covering the syllabus and makes it impossible for you to avoid those topics either which you do not like or which you find difficult without incurring a loss of marks in the examination considered as a whole.

Some examination boards achieve the same effect by setting questions of variable length with a mark allocation shown at the side or end of each question. In such papers the total mark allocation normally exceeds that required for full marks by about 10 per cent.

Mark schemes and solutions for each question in Papers 5, 6, 8 and 9 are provided for you. We suggest you try the papers before looking at the solutions and mark allocations, then mark your own attempts and see how well, or badly, you have performed. Learn from your mistakes, particularly if you have made errors of principle, but also notice how errors in arithmetic and/or manipulation can lead you astray. Try the papers again a few weeks later and test yourself to see how much you have improved and how many mistakes have been avoided second time round. Some short questions tests are given in Chapters 8 and 9.

(iii) *Long questions*

Generally, you are given a choice and this choice must be exercised wisely. Select the question you consider to be the best for you, read all of it, and answer this one first. Remember that these questions tend to be carefully structured and that hints are often given by the examiner to help you along.

2 Mathematical thinking

2.1 Pure mathematics

Essentially, the process of thinking consists of asking yourself questions—the right questions of course! Consider, for example, the following question from a pure mathematics paper.

Example 1 Find the first four terms in the series expansion of $\dfrac{1}{3x + 5}$ in ascending powers of x, and state the set of values of x for which this expansion is valid.

What questions do you ask yourself?

(i) Have I seen this question, or one like it, before? If so, how was it done?

(ii) On what part of the syllabus is it based? [The binomial theorem.] But the binomial theorem is, for $n \notin \mathbb{N}$ and $|X| < 1$,

$$(1 + X)^n = 1 + nX + \frac{n(n-1)}{2!} X^2 + \frac{n(n-1)(n-2)}{3!} X^3 + \dots .$$

(iii) Then how can I transform $\dfrac{1}{3x + 5}$ into the form $(1 + X)^n$? By writing

$$\frac{1}{(3x + 5)} = \frac{1}{5(1 + 3x/5)} = \tfrac{1}{5}(1 + 3x/5)^{-1},$$

you find, using $n = -1$, $X = 3x/5$,

$$\frac{1}{3x + 5} = \frac{1}{5}\left[1 + (-1)\frac{3x}{5} + \frac{(-1)(-1-1)}{2!}\left(\frac{3x}{5}\right)^2 \right.$$
$$\left. + \frac{(-1)(-1-1)(-1-2)}{3!}\left(\frac{3x}{5}\right)^3 + \dots \right].$$

Note that the question asks for only the first four terms in the expansion; this is as far as you go. [*Do not do unnecessary work.*] Then, simplifying,

$$\frac{1}{3x + 5} = \frac{1}{5} - \frac{3x}{25} + \frac{9x^2}{125} - \frac{27x^3}{625} + \dots .$$

[Quickly check the arithmetic.]

(iv) Have I finished the question? [Read it through and notice that the set of values of x for which the expansion is valid is required.]

(v) Since the binomial expansion of $(1 + X)^n$ is valid for $|X| < 1$, then our expansion is valid for $|3x/5| < 1 \Leftrightarrow |x| < 5/3$.

The set of values of x is $\{x : -5/3 < x < 5/3\}$.

Example 2 Consider the evaluation of $\displaystyle\int_0^1 x\,e^{-2x}\,dx$.

(i) How can I work out integrals such as $\displaystyle\int x\,e^{-2x}\,dx$?

Is it a standard form or can it be reduced easily to one? [No.]

What techniques of integration do I know? [Substitution, use of partial fraction, parts.]

I note that we have here the product of an algebraic function and an exponential function and such integrals are almost *always* done by the method of 'integration by parts'.

$$\frac{d}{dx}(e^{-2x}) = -2\,e^{-2x}$$

and so write

$$\int_0^1 x\,e^{-2x}\,dx = \int_0^1 x\,\frac{d}{dx}\left(\frac{-e^{-2x}}{2}\right)dx$$

$$= \left[x\,\frac{(-e^{-2x})}{2}\right]_0^1 - \int_0^1 \frac{(-e^{-2x})}{2}\,dx$$

$$= -\tfrac{1}{2}e^{-2} + \frac{1}{2}\int_0^1 e^{-2x}\,dx$$

$$= -\tfrac{1}{2}e^{-2} + \frac{1}{4}\left[e^{-2x}\right]_0^1$$

$$= -\tfrac{1}{2}e^{-2} - \tfrac{1}{4}e^{-2} + \tfrac{1}{4}e^0$$

$$= \tfrac{1}{4} - \tfrac{3}{4}e^{-2}.$$

(ii) This is the answer—you need not find the numerical value ($\approx 0 \cdot 15$) in this case unless you are told to give it to so many decimal places or significant figures.

(iii) Is the answer reasonable? [It looks like it, since $x\,e^{-2x}$ has a maximum $\tfrac{1}{2}e^{-1}$ ($\approx 0 \cdot 18$) when $x = \tfrac{1}{2}$ and, when $x = 1$, $x\,e^{-2x} = e^{-2}$ ($0 \cdot 135$), although you need not carry out such detailed calculations.]

Example 3 Consider the question:

Find the solution set of the inequation

$$\frac{x}{x+6} > \frac{1}{x+2}.$$

(i) How do I solve such problems? [By multiplication by $(x + 6)(x + 2)$ and so clearing of fractions? No—this is only valid, when dealing with an inequation such as this, if $(x + 6)(x + 2) > 0$ and this may not be true.] Remember that it is easier to solve the inequation $f(x) > 0$ than the inequation $\phi(x) > \psi(x)$.

(ii) Therefore, consider

$$\frac{x}{x + 6} - \frac{1}{x + 2} > 0.$$

What do I do now? [Express the left-hand side as a single fraction.]

$$\frac{x(x + 2) - (x + 6)}{(x + 6)(x + 2)} > 0$$

$$\Rightarrow \frac{x^2 + x - 6}{(x + 6)(x + 2)} > 0.$$

(iii) What now? [Factorize the numerator. This is usually possible in this type of question.]

$$\frac{(x - 2)(x + 3)}{(x + 6)(x + 2)} > 0.$$

(iv) How can I proceed? [Consider the signs of the various factors in the numerator and denominator.]

Value of x	$x < -6$	$-6 < x < -3$	$-3 < x < -2$	$-2 < x < 2$	$2 < x$
Sign of $x + 6$	$(-)$	$(+)$	$(+)$	$(+)$	$(+)$
Sign of $x + 3$	$(-)$	$(-)$	$(+)$	$(+)$	$(+)$
Sign of $x + 2$	$(-)$	$(-)$	$(-)$	$(+)$	$(+)$
Sign of $x - 2$	$(-)$	$(-)$	$(-)$	$(-)$	$(+)$
Sign of $\dfrac{(x - 2)(x + 3)}{(x + 6)(x + 2)}$	$(+)$	$(-)$	$(+)$	$(-)$	$(+)$

Solution set $x < -6$ and $-3 < x < -2$ and $2 < x$;

or $\qquad \{x : x < -6\} \cup \{x : -3 < x < -2\} \cup \{x : 2 < x\}.$

(v) Is the result reasonable? [Check when $x = -10, -4, -2\frac{1}{2}, 0, 4$. It looks correct.]

Example 4 Consider the question:

Given that $f(x) = \dfrac{1}{x^2 - 2x + 3}$, show that $f(x)$ is always positive.

Sketch the graph of $y = f(x)$, stating the equations of the asymptotes and the coordinates of the points of intersection (if any) with the coordinate axes.

(i) Have I seen a question of this type before and, if so, how was it done?

(ii) On what part of the syllabus is it based? Noting the form of the denominator, the answer is quadratic theory. To see how this idea will help, we observe that $f(x) > 0$ if $x^2 - 2x + 3 > 0$.

(iii) In dealing with the sign of a quadratic expression we either factorize or express as a sum of squares. An inspection of $x^2 - 2x + 3$ shows us that there are no real factors, in which case we should be able to express as a sum of squares. This is achieved by 'completing the square' on $x^2 - 2x$, i.e. $(x - 1)^2$, hence

$$x^2 - 2x + 3 \equiv (x - 1)^2 + 2.$$

(iv) We have now arrived at

$$f(x) = \frac{1}{(x - 1)^2 + 2}.$$

But the sum of squares is positive for all real values of x and so $f(x)$ is positive for all real values of x.

(v) How to 'sketch' a graph? By noting its important features, such as restrictions on the values of x or y, asymptotes, intersections with the axes, and turning points. Always let the question help you, as it will contain some guidance on which of these points to look out for. In this question you were asked to prove $f(x)$ always positive, i.e. $y > 0$, which cuts out half of the x–y plane. Now look at the other points specifically mentioned.

(vi) Asymptotes. Any values of x for which $y \to \infty$? No, as $x^2 - 2x + 3 \neq 0$. So there are no asymptotes parallel to the y-axis. Next, let $x \to \pm \infty$, for which $y \to 0$, so that $y = 0$ is an asymptote.

(vii) Curve cuts the y-axis at $(0, \frac{1}{3})$; no intersection with the x-axis.

(viii) Turning points. $(x - 1)^2 + 2$ has minimum value 2 when $x = 1$. Hence $f(x)$ has maximum value $\frac{1}{2}$ when $x = 1$. We now have sufficient to sketch the graph. [It is often a help to 'shade out' the part of the x–y plane in which there are no values for x or y.]

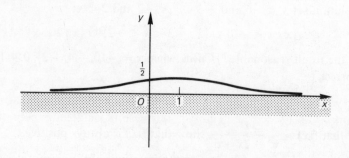

Note that you should try not to get involved in complicated differentiation to find maxima or minima. They can often be easily found algebraically, as above.

Be flexible in your approach. For instance, the use of vector ideas can often simplify coordinate geometry questions.

Example 5   Consider the simple question (which could be part of a longer one):
Given the points A $(1, 2)$, B $(2, 3)$, C $(5, 1)$, find the coordinates of D such that $ABCD$ is a parallelogram.

(i) How can I make use of the fact that AB is equal and parallel to DC? Instead of getting involved in coordinate geometry write

$$\overrightarrow{AB} = \overrightarrow{DC}. \tag{1}$$

(ii) Let D be the point (x, y) and then (1) may be written

$$\begin{pmatrix} 2 \\ 3 \end{pmatrix} - \begin{pmatrix} 1 \\ 2 \end{pmatrix} = \begin{pmatrix} 5 \\ 1 \end{pmatrix} - \begin{pmatrix} x \\ y \end{pmatrix}$$

$$\Leftrightarrow \begin{pmatrix} 1 \\ 1 \end{pmatrix} = \begin{pmatrix} 5 - x \\ 1 - y \end{pmatrix}$$

$$\Leftrightarrow \quad x = 4, \qquad y = 0,$$

i.e. D is the point $(4, 0)$.

A further vector problem follows:

Example 6   The two lines with vector equations

$$\mathbf{r} = \mathbf{i} + \lambda(\mathbf{j} + \mathbf{k}), \qquad \mathbf{r} = \mathbf{i} + \mu(\mathbf{j} - \mathbf{i})$$

intersect at the point A. Write down the position vector of A.
   Find a vector perpendicular to both lines and, hence or otherwise, obtain a vector equation of the plane containing the two lines.

(i) How do I find the intersection point of two lines?
   The vector equations of the lines give us, in each case, the position vector of any point on the line, so the point A which lies on both lines is given by

$$\mathbf{i} + \lambda(\mathbf{j} + \mathbf{k}) = \mathbf{i} + \mu(\mathbf{j} - \mathbf{i})$$

$$\Leftrightarrow \quad \mathbf{i} + \lambda\mathbf{j} + \lambda\mathbf{k} = (1 - \mu)\mathbf{i} + \mu\mathbf{j}.$$

It is easily seen that this equation is satisfied by $\lambda = \mu = 0$ and A is the point with position vector $\mathbf{i}$.
   (When a question says 'write down' it is an indication that no real working is needed and, indeed, this result could have been seen by inspection.)

(ii) Can I write down a vector perpendicular to both lines? Not directly, but suppose such a vector is $(a\mathbf{i} + b\mathbf{j} + c\mathbf{k})$.
   I have to express that this vector is perpendicular to the directions of the given lines. These directions are $(\mathbf{j} + \mathbf{k})$ and $(\mathbf{j} - \mathbf{i})$.

(iii) How do I express the fact that two vectors are perpendicular? By writing down their scalar products and equating those products to zero. Thus

$$(a\mathbf{i} + b\mathbf{j} + c\mathbf{k}) \cdot (\mathbf{j} + \mathbf{k}) = 0, \qquad (a\mathbf{i} + b\mathbf{j} + c\mathbf{k}) \cdot (\mathbf{j} - \mathbf{i}) = 0$$

$$\Rightarrow \quad b + c = 0, \qquad -a + b = 0$$

$$\Rightarrow \quad a = b = -c.$$

The perpendicular vector is, therefore, $(a\mathbf{i} + a\mathbf{j} - a\mathbf{k})$ or $(\mathbf{i} + \mathbf{j} - \mathbf{k})$.

(iv) Finally, for the equation of the plane we note that we have already found a point on the plane and the direction of the normal to the plane and so we can write down an equation of the plane as

$$(\mathbf{r} - \mathbf{i}) \cdot (\mathbf{i} + \mathbf{j} - \mathbf{k}) = 0.$$

Example 7 Find the sum $\displaystyle\sum_{r=2}^{n} \frac{1}{(r-1)(r+1)}$.

(i) Have I seen a question like this before and how was it done? The expression suggests partial fractions, so start with

$$\frac{1}{(r-1)(r+1)} = \frac{1}{2(r-1)} - \frac{1}{2(r+1)}$$

and so

$$\sum_{r=2}^{n} \frac{1}{(r-1)(r+1)} = \sum_{r=2}^{n} \frac{1}{2(r-1)} - \sum_{r=2}^{n} \frac{1}{2(r+1)}.$$

Note that it is a good idea to keep a list of all the different types of questions in which partial fractions are helpful.

(ii) As each term of the series is now expressed as a difference we would expect most terms to 'cancel out' when the series is expanded. Thus

$$\sum_{r=2}^{n} \frac{1}{(r-1)(r+1)} = \frac{1}{2}\left[\left(1 - \frac{1}{3}\right) + \left(\frac{1}{2} - \frac{1}{4}\right) + \left(\frac{1}{3} - \frac{1}{5}\right) + \ldots\right.$$

$$\left. + \left(\frac{1}{n-2} - \frac{1}{n}\right) + \left(\frac{1}{n-1} - \frac{1}{n+1}\right)\right]$$

$$= \frac{1}{2}\left[\frac{3}{2} - \frac{1}{n} - \frac{1}{n+1}\right].$$

This may be more neatly expressed as

$$\frac{1}{2}\sum_{r=1}^{n} \frac{1}{r-1} - \frac{1}{2}\sum_{r=2}^{n} \frac{1}{r+1} = \frac{1}{2}\sum_{r=2}^{n} \frac{1}{r-1} - \frac{1}{2}\sum_{r=4}^{n+2} \frac{1}{r-1}$$

$$= \frac{1}{2}\left[\frac{3}{2} - \frac{1}{n} - \frac{1}{n+1}\right].$$

2.2 Applied mathematics

Example 1   A particle, of mass m, lies on a horizontal platform which is being accelerated vertically upwards with an acceleration of magnitude f. Find the magnitude of the force exerted by the platform on the particle.

What questions do you ask yourself?

(i) Should I draw a figure and mark in it the forces acting on the particle and the acceleration of the particle? [Of course.]

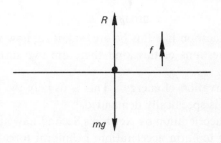

(ii) Have I seen this question, or one like it, before? If so, how was it done?

(iii) On what part of the syllabus is it based? [Using Newton's Second Law in the form, force = mass × acceleration.]

(iv) Write down the vertically upwards component of the equation of motion of the particle:

$$R - mg = mf$$
$$\Rightarrow \quad R = m(g + f).$$

(v) Have I included all the forces and put them and the acceleration in the correct directions? [Yes.]

(vi) Is the answer reasonable? For example, are the dimensions correct? In any equation of theoretical mechanics the dimensions of each term should be the same. [The dimensions agree.]

In applied mathematics, consideration of this last question is most important. Examiners try to set questions with sensible answers and you should always bear this in mind.

For example, consider the question:

Example 2   A car, of mass 2000 kg, moving at 36 km h^{-1} along a straight level road, has the brakes suddenly applied and the engine shut off. Given that the resistance to motion is of magnitude 5000 N, find the distance covered by the car before it comes to rest.

(i) Should I draw a figure and mark on it the forces acting on the car? [Yes. Although a figure is not strictly necessary here, many people find that it clarifies matters for them and it need only be a small freehand diagram.]

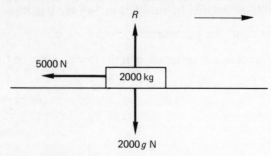

(ii) Have I seen a question like this before and, if so, how was it done?

For many questions of this kind there are two distinct methods of approach.

(a) Using conservation of energy. [This is usually the easier unless the acceleration is specifically demanded.]

(b) Finding the acceleration by Newton's Second Law and then using the equations of uniform acceleration. [Uniform force implies uniform acceleration.]

Considering (a): In general we say

Initial (KE + PE) − work done against resistance = final (KE + PE).

Initial speed is $36 \times \frac{5}{18}$ m s^{-1} = 10 m s^{-1}. [Why do we need to convert to m s^{-1}?] Initial KE = $\frac{1}{2}$. 2000 . 10^2 J. PE zero throughout. [Why?] Work done against resistance = 5000s J (where s m is distance covered). Hence

$$\tfrac{1}{2} . 2000 . 10^2 - 5000s = 0$$

$$\Rightarrow \quad s = 20.$$

Considering (b): Using

$$F = ma, \qquad -5000 = 2000a \quad \Rightarrow \quad a = -2.5,$$

i.e. the acceleration is -2.5 m s^{-2} (a retardation).

Initial speed = 10 m s^{-1} as in (a).
Final speed = 0 m s^{-1} as car comes to rest.

In formulae for constant acceleration we have u, v, a and need to find s. Hence use

$$v^2 = u^2 + 2as$$

$$\Rightarrow \quad s = \frac{100}{2 \times 2.5}$$

$$= 20$$

as in (a).

(iii) Thus we have that the distance covered is 20 m. This seems a reasonable answer. If we had found the answer to be 2000 m (2 km) or 20 cm, common sense would suggest that we had an error of some kind, a g introduced unnecessarily perhaps?

Now consider the following question:

Example 3   A uniform rod AB rests, with the end A against a smooth vertical wall, in a vertical plane perpendicular to the wall with the end B on a rough horizontal plane. Given that the coefficient of friction between the rod and the ground is μ, find the set of possible values of θ, the inclination of the rod to the vertical.

(i) What type of problem is this? [Clearly a statics problem involving friction.]
(ii) First draw a figure, marking in all the forces acting on the rod:

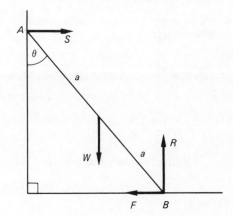

(iii) Have we marked in every force (or component) and the remainder of the given data? [Yes; check why this figure is correct.]
   Note that we must not assume limiting friction unless the question specifically states that the friction is limiting or that this is implied by the use of expressions such as 'about to slip'. Equilibrium can be maintained even if friction is less than limiting.
(iv) What is the eventual target? [To use the inequality $F/R \leqslant \mu$ and so obtain an inequality involving θ.]
(v) How many unknowns are there? [Three, namely F, R and S.]
(vi) How many equations do I need to determine F and R (or better F/R)? [Usually we need three equations for three unknowns, though this requirement is sometimes reduced when a ratio only is required.]
(vii) How are three equations found? [By resolving in two perpendicular

directions and taking moments about a fixed point. This is by far the most usual method, although there are statics questions in which it may be easier to take more than one set of moments.]

Thus:

$$\text{Resolving vertically,} \qquad R = W. \qquad (1)$$

$$\text{Resolving horizontally,} \qquad F = S. \qquad (2)$$

About which point should moments be taken? [Bearing in mind that a force does not occur in a moments equation when its line of action passes through the point chosen, we usually choose the point which has most lines of action passing through it, thus cutting out the maximum number of forces and so simplifying the subsequent algebra.]

In this case then we choose B.

Taking moments about B,

$$Wa \sin \theta = 2Sa \cos \theta \qquad \text{(where } 2a = \text{length of rod)}$$

$$\Rightarrow \quad W \sin \theta = 2S \cos \theta \qquad (3)$$

$$\Rightarrow \quad \tan \theta = \frac{2S}{W}$$

$$\Rightarrow \quad \tan \theta = \frac{2F}{R} \qquad \text{(using equations (1) and (2))}$$

$$\Rightarrow \quad \frac{F}{R} = \tfrac{1}{2} \tan \theta.$$

But $$\qquad F/R \leqslant \mu \quad \Rightarrow \quad \tan \theta \leqslant 2\mu$$

$$\Rightarrow \qquad \theta \leqslant \tan^{-1}(2\mu).$$

Also $\theta \geqslant 0$. [Why?]

So our result is

$$0 \leqslant \theta \leqslant \tan^{-1}(2\mu).$$

Note that we could have managed without equation (2) if we had chosen to take moments about A, but we should then have lost the advantage of the very easy equation (3).

Example 4 A train takes two minutes to travel from station A to station B, a distance of 2400 m. The train accelerates uniformly from rest at A to a maximum speed of 108 km h^{-1}, travels at that maximum speed and then decelerates uniformly to rest at B. The time spent decelerating is three times that spent accelerating. Find the acceleration.

(i) Have I seen this type of problem before? What part of the syllabus is it on? [Uniform acceleration.]

(ii) What method should be used? [There are two distinct methods,
   (a) by drawing a speed–time graph—a sketch, not an accurate drawing,
   (b) by using the equations of uniform acceleration.
   The former is often the quicker method for this type of problem.]
(iii) For either method we need to write maximum speed $= 108 \times \frac{5}{18}$ m s^{-1}
   $= 30$ m s^{-1}. [Why?]

Considering method (a):
(1) Drawing the sketch. This will consist of straight lines. [Why?]

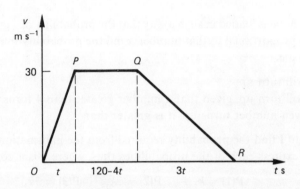

(2) To make use of the data concerning time, we put time accelerating
   $= t$ seconds, then time decelerating $= 3t$ seconds and time at uniform
   speed $= (120 - 4t)$ seconds. (Using the fact that the journey takes two
   minutes.)
(3) How do we find the required acceleration? [The acceleration is the
   gradient of the part OP of the graph. Hence we need to find t in order to
   determine this gradient.]
(4) Refer to the question again to see whether there are any data not yet
   used. We are given the distance between the stations and the distance in
   a speed–time graph is the area under the curve. Hence

$$\tfrac{1}{2} \cdot 30t + 30(120 - 4t) + \tfrac{1}{2} \cdot 30 \cdot 3t = 2400$$

$$\Rightarrow \quad 3600 - 60t = 2400 \quad \Rightarrow \quad t = 20.$$

Acceleration $=$ gradient of $OP = 30/20$ m s$^{-2} = 1 \cdot 5$ m s^{-2}.

Considering method (b):
(1) As for (2) in (a).
(2) How do we find the acceleration? Again, we need to find the time spent
   accelerating, so that we need to use the fact that the total distance for
   the journey is 2400 m.
(3) Using the formula $s = \dfrac{(u + v)t}{2}$:

Distance covered accelerating $= \frac{1}{2} \times 30t$ m.

Distance covered at uniform speed $= 30 \times (120 - 4t)$ m.

Distance covered decelerating $= \frac{1}{2} \times 30 \times 3t$ m.

$$\text{Total distance} = \left[15t + 30(120 - 4t) + 45t\right] \text{m}$$

$$= 2400 \text{ m} \quad \Rightarrow \quad t = 20 \qquad \text{as in (a).}$$

(4) For the period spent accelerating we have u, v and t and we need to find a, so we use the formula $v = u + at$;

$$\Rightarrow \quad 30 = 20a \quad \Rightarrow \quad a = 1 \cdot 5 \qquad \text{as in (a).}$$

Example 5 A die is loaded in such a way that the probability of a given number turning up is proportional to that number. Find the probability that, in a single throw of the die:

(a) a six will turn up;

(b) a six will turn up, given that a number greater than 4 turns up;

(c) if an even number turns up, it is greater than 3.

(i) How do I find the probability required from the information given? We could express each of the probabilities thus, where k is a constant,

$$P(1) = k, \qquad P(2) = 2k, \qquad P(3) = 3k,$$

$$P(4) = 4k, \qquad P(5) = 5k, \qquad P(6) = 6k.$$

(ii) Can we find k? [We make use of the fact that the sum of all these probabilities is 1 (the list is exhaustive).] Hence

$$21k = 1 \quad \Rightarrow \quad k = 1/21.$$

(iii) This gives $P(6) = 6/21 = 2/7$.

(iv) How does this probability differ from the preceding one? [It is an example of conditional probability.]

We need to take the two events separately.

$$P(6) = 2/7, \qquad P(\text{number greater than 4}) = 5k + 6k$$

$$= 11/21$$

$$\Rightarrow \quad P(6 \,|\, \text{number} > 4) = \frac{2}{7} \bigg/ \frac{11}{21} = 6/11.$$

(v) (c) is also a conditional probability, but this time we need to consider which of the even numbers are greater than 3, i.e. 4 and 6. Then

$$P(4 \text{ or } 6) = 4k + 6k = 10/21, \ P(\text{even number}) = 2k + 4k + 6k = 12/21$$

$$\Rightarrow \quad P(\text{number} > 3 \,|\, \text{even number}) = \frac{10}{21} \bigg/ \frac{12}{21} = 5/6.$$

Example 6 Ship A has constant velocity of magnitude 10 m s^{-1} due east. Relative to A, ship B has constant velocity of magnitude 16 m s^{-1} bearing $210°$. Find the velocity of B as recorded by an observer at a nearby lighthouse.

At a particular instant, B is 2000 m due east of A. Find, in minutes, the time taken from this instant until the distance between the ships is least. Find also this least distance between them.

In questions on relative velocity a clear understanding of the principles involved is of the utmost importance and the use of vector methods aids both the understanding and the working out of problems.

(i) The given problem is not set in vector form but can it be converted to this form? In questions such as this, involving bearings, by taking the unit vectors $\mathbf{i}$ and $\mathbf{j}$ in the east and north directions respectively the problem can be put in vector form.

Thus, if $\mathbf{V}_A \text{ m s}^{-1}$ and $\mathbf{V}_B \text{ m s}^{-1}$ are the velocities of ships A and B respectively,

$$\mathbf{V}_A = 10\mathbf{i}. \tag{1}$$

And, as the velocity of B relative to A is $(\mathbf{V}_B - \mathbf{V}_A) \text{ m s}^{-1}$,

$$\mathbf{V}_B - \mathbf{V}_A = -8\mathbf{i} - 8\sqrt{3}\mathbf{j}. \tag{2}$$

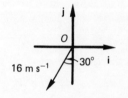

(a small diagram enables us to see this).
From (1) and (2),

$$\mathbf{V}_B = 2\mathbf{i} - 8\sqrt{3}\mathbf{j},$$

which answers the first part of the question, though it should be put in the form in which the velocities are given in the question, i.e.

$$|\mathbf{V}_B| = \sqrt{(4 + 192)} \text{ m s}^{-1} = 14 \text{ m s}^{-1};$$

angle with south is

$$\tan^{-1}\left[2/(8\sqrt{3})\right] \approx 8\cdot2°.$$

Thus the velocity of B is 14 m s^{-1} on a bearing of $171\cdot8°$.

(ii) How do I find the least distance between the ships? It is possible to find the relative distance in terms of the time from integration of (2) and then find when this distance is a minimum, but this can involve quite heavy working and the problem can be solved more simply. The fact that the velocity of B relative to A is 16 m s^{-1} bearing $210°$ means that we can regard A as stationary and B as moving at this velocity having started due east of A.

Hence the diagram.

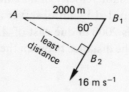

Least distance $= AB_2 = 2000 \sin 60°$ m ≈ 1732 m.

Also, $B_1B_2 = 2000 \cos 60°$ m $= 1000$ m and this distance is covered at the *relative* speed, i.e. 16 m s^{-1}, and so the time $= 1000/16$ s $\approx 1\cdot04$ minutes.

3 Marking

3.1 General principles

How are marks awarded for attempts at mathematics questions? Three main types of marks are given. They are method marks (**M**), accuracy marks (**A**), follow through marks (**FT**).

Method marks (**M**) are given for knowing what to do in a particular case, i.e. recognizing the principles and techniques required, *and* attempting to carry out the manipulative process required. To earn these marks you must attempt to solve the problem mathematically. An explanation of what you are trying to do, such as 'using the substitution $x = a \sin \theta$', 'using the sine rule for $\triangle ABC$', 'using the remainder theorem', 'interchanging the roles of x and y', 'separating the variables', 'using Newton's Second Law', 'taking moments about A $(A\,\widehat{\,})$', 'integrating this differential equation', 'multiplying these independent probabilities', will help in gaining method marks. However, as previously stated, a verbal offering without the benefit of mathematical equations will receive no credit. The examiner knows how to do the question and his/her duty is to find out whether *you* know how to do it, and correctly at that.

Accuracy marks (**A**) are given for correct working, *after the method marks have been earned*. No accuracy marks can be awarded until the appropriate method marks have been earned. However, method marks can be earned even if numerical and/or algebraic errors lead to loss of accuracy marks.

Follow through marks (**FT**) are accuracy marks given for correct working after a numerical or algebraic error (from the candidate's own mistake). You could make an error early in a question but still gain almost all the marks provided your calculations, after an error, follow on correctly from your mistake. But you must have the method correct! Further, if intermediate answers are printed on the paper, you must use these in your subsequent work—examiners will *not* follow through when you ignore correct results given to you as a help and check on your work.

There are other marks, sometimes referred to as **B** marks, which are usually regarded as accuracy marks not dependent on method marks.

Correct answer only marks (**cao**) are those accuracy marks, usually towards the end of a question or part question, which are only awarded for a correct answer following on correct previous work. So these are only given for fully correct work.

3.2 Pure mathematics example

Example As an example of marking procedure, we will consider the inequation of Example 3, §2.1, and give three solutions, one correct and two containing mistakes, together with the appropriate marks earned.

Solution 1. Correct.

$$\frac{x}{x+6} - \frac{1}{x+2} > 0 \qquad \textbf{(M1)}$$

$$\Rightarrow \quad \frac{x(x+2) - (x+6)}{(x+6)(x+2)} > 0 \qquad \textbf{(M2 A1)}$$

$$\Rightarrow \quad \frac{x^2 + x - 6}{(x+6)(x+2)} > 0 \qquad \textbf{(A1 FT)}$$

$$\Rightarrow \quad \frac{(x-2)(x+3)}{(x+6)(x+2)} > 0. \qquad \textbf{(A1 cao)}$$

For the table in the example (p. 9) would be given (**M1 A2, 1** or **0 FT**, one **A** mark lost for each of first two **A** errors).

The final correct statement of the answer, in either form. **(A1 cao)**

(Total 10 marks—full marks)

Solution 2. Incorrect. [Errors shown in square boxes.]

$$\Rightarrow \quad \frac{x}{x+6} - \frac{1}{x+2} > 0 \qquad \textbf{(M1)}$$

$$\Rightarrow \quad \frac{x(x+2) \;\boxed{+}\; (x+6)}{(x+6)(x+2)} > 0 \qquad \textbf{(M2 A0)}$$

$$\Rightarrow \quad \frac{x^2 + 3x + 6}{(x+6)(x+2)} > 0 \qquad \textbf{(A1 FT)}$$

$$\Rightarrow \quad \frac{(x-2)(x-3)}{(x+6)(x+2)} > 0. \qquad \textbf{(A0 cao)}$$

Values of x	$x < -6$	$-6 < x < -2$	$-2 < x < 2$	$2 < x < 3$	$x > 3$
Sign of $x + 6$	$(-)$	$(+)$	$(+)$	$(+)$	$(+)$
Sign of $x + 2$	$(-)$	$(-)$	$(+)$	$(+)$	$(+)$
Sign of $x - 2$	$(-)$	$(-)$	$(+)$	$(+)$	$(+)$
Sign of $x - 3$	$(-)$	$(-)$	$(-)$	$(-)$	$(+)$
Sign of $\dfrac{(x-2)(x-3)}{(x+6)(x+2)}$	$(+)$	$(-)$	$(+)$	$\boxed{(+)}$	$(+)$

Table **(M1 A1 FT)**

Answer: $\{x : x < -6\} \cup \{x : x > -2\}.$ **(A0 cao)**

(Total 6 marks)

Solution 3. Incorrect. [Error given in rectangle.]

| Multiply by $(x + 6)(x + 2)$ | **(M0)**

$\Rightarrow \quad x(x + 2) > x + 6$

$\Rightarrow \quad x^2 + x - 6 > 0$

$\Rightarrow \quad (x - 3)(x + 2) > 0$

$\Rightarrow \quad \{x : x < -2\} \cup \{x : x > 3\}.$

(Despite the fact that these lines follow on from one another, no marks are given because of **M0**—following hopeless starting error of principle.)

(Total 0 marks)

3.3 Applied mathematics examples

Example 1 As an example of marking procedure, we consider the statics question which was solved in Chapter 2 (Example 3, § 2.2). We give three solutions, one correct and two containing errors, together with the appropriate marks earned.

Solution 1. Correct.

Resolving vertically, $\quad R = W.$ (1) **(M1)**

Resolving horizontally, $\quad F = S.$ (2) **(M1)**

$B \circlearrowright \quad Wa \sin \theta = 2Sa \cos \theta$ **(M2 A1)**

$\Rightarrow \quad \tan \theta = \dfrac{2S}{W}$ (3) **(A1)**

$\Rightarrow \quad \dfrac{F}{R} = \frac{1}{2} \tan \theta \quad$ (using 1 and 2). **(A1)**

But $\quad \dfrac{F}{R} \leqslant \mu$ **(M2)**

$\Rightarrow \quad \theta \leqslant \tan^{-1}(2\mu).$ **(A1 cao)**

Also $\quad \theta \geqslant 0 \quad$ and so

$0 \leqslant \theta \leqslant \tan^{-1}(2\mu).$ **(A1 cao)**

(Total 11 marks—full marks)

Notice that six marks are given for use of the correct principles. About half marks are given in applied mathematics for method, but A(**cao**) marks are common.

Solution 2. Incorrect. Errors are given in square boxes.
(Note that mark schemes are adapted to take account of different methods. The two **M** marks allowed for the resolutions will both go to the one resolution for the candidate who sees that only one is necessary when moments are taken about A.)

Resolving vertically, $\quad R = W.$ (1) **(M2)**

$A \circlearrowright \quad R \cdot 2a \boxed{\cos \theta} = F \cdot 2a \boxed{\sin \theta} + Wa \boxed{\cos \theta}.$ (2) **(M2 A0)**

Using (1) and (2) $\Rightarrow$ $F = \frac{1}{2}R \cot \theta$ **(A1 FT)**

$\Rightarrow$ $F/R = \frac{1}{2} \cot \theta$.

But $F/R \leqslant \mu$ **(M2)**

$\Rightarrow$ $\theta \leqslant \cot^{-1}(2\mu)$. **(A0 cao)**

Solution ends with no consideration of $\theta \geqslant 0$. **(Total 7 marks)**

Solution 3. Very incorrect. Errors are given in square boxes.

$A \curvearrowright$ $F . 2a \boxed{\cos \theta} = R . 2a \boxed{\sin \theta}$ (*W* omitted) **(M2 A0)**

$\Rightarrow$ $F/R = W \tan \theta$. **(M0)**

But $\boxed{F/R = \mu}$ $\Rightarrow$ $\mu = \tan \theta$. **(M0)**

No more working. **(Total 2 marks)**

Example 2 As a further example of marking procedure, we consider the probability question solved in Chapter 2 (Example 5, § 2.2). We give three solutions, one correct and two containing errors, together with the appropriate marks earned.

Solution 1. Correct.

$$P(1) = k, \qquad P(2) = 2k, \qquad P(3) = 3k,$$
$$P(4) = 4k, \qquad P(5) = 5k, \qquad P(6) = 6k.$$

(M2)

Adding,

$$21k = 1 \quad \Rightarrow \quad k = 1/21. \qquad \textbf{(M1 A1)}$$

(a) $P(6) = 6k = 2/7$. **(A1 cao)**

(b) $P(\text{number} > 4) = 5k + 6k = 11/21$. **(M1 A1)**

$$P(6 \,|\, \text{number} > 4) = \frac{2}{7} \bigg/ \frac{11}{21} = 6/11. \qquad \textbf{(M1 A1)}$$

(c) Even numbers > 3 are 4 and 6,

$$P(4 \text{ or } 6) = 4k + 6k = 10/21. \qquad \textbf{(M1 A1)}$$

$$P(\text{even number}) = 2k + 4k + 6k = 12/21 \qquad \textbf{(M1 A1)}$$

$\Rightarrow$ $P(\text{number} > 3 \,|\, \text{even number}) = \dfrac{10}{21} \bigg/ \dfrac{12}{21} = 5/6.$ **(M1 A1)**

(Total 15 marks—full marks)

Solution 2. Incorrect. Errors given in boxes.

$$P(1) = k, \qquad P(2) = 2k, \qquad P(3) = 3k,$$
$$P(4) = 4k, \qquad P(5) = 5k, \qquad P(6) = 6k.$$

(M2)

Adding,
$$21k = 1 \quad \Rightarrow \quad k = 1/21.$$ **(M1 A1)**

(a) P(6) = 2/7. **(A1)**

(b) P(number > 4) = 11/21. **(M1 A1)**

P(6 | number > 4) = $\boxed{2/7 \times 11/21}$ = 22/147. **(M0 A0)**

(c) Even numbers > 3 are 4, 6,
$$P(4, 6) = 4k + 6k = 10/21.$$ **(M1 A1)**
$$P(\text{even number}) = 2k + 4k + 6k = 12/21.$$ **(M1 A1)**
$$\Rightarrow \quad P(\text{number} > 3 \,|\, \text{even number}) = \boxed{\frac{10}{21} \times \frac{12}{21}} = \frac{40}{147}.$$ **(M0 A0)**

(Total 11 marks)

Solution 3. Incorrect. Errors are given in boxes.

$$P(1) = \boxed{1/36}, \qquad P(2) = \boxed{2/36}, \qquad P(3) = \boxed{3/36},$$
$$P(4) = \boxed{4/36}, \qquad P(5) = \boxed{5/36}, \qquad P(6) = \boxed{6/36}.$$ **(M1)**

(M0 A0)

(The only correct part of this is getting probabilities proportional to scores.)

(a) P(6) = 1/6. **(A0 cao)**

(b) P(number greater than 4) = $\boxed{11/36}$. **(M1 A0)**

$$P(6 \,|\, \text{number} > 4) = \frac{1}{6} \bigg/ \frac{11}{36} = 6/11.$$ **(M1 A1)**

(c) $\qquad\qquad\qquad$ P(even number) = 12/36. **(M1 A0)**

$$\boxed{P(\text{number} > 3) = 15/36}$$ **(M0 A0)**

$$\Rightarrow \quad \boxed{P(\text{number} > 3 \,|\, \text{even number}) = \frac{12}{36} \bigg/ \frac{15}{36}.}$$ **(M0 A0)**

(The **A** mark in (b) may appear to be rather luckily gained but it could fairly be considered as an **FT** mark.)

(Total 5 marks)

Some examination boards adopt the following convention on equations in mechanics. The requirement for the award of any **M** marks is that all relevant forces or terms for the equation should be present and no additional forces or terms. Also, all terms should be dimensionally correct. Sin for cos and vice versa are treated as **A** errors but the omission of sin or cos where required gets **M0**, e.g. W for $W \sin \alpha$ or $W \cos \alpha$.

In some of the illustrative papers of Chapters 6 and 7, and the examples of Chapters 8–10, mark schemes are shown in detail. If you try the papers before looking at the solutions and explanations, mark your own attempts and see how well, or badly, you have performed. Learn from your mistakes, particularly if you made errors of principle, but notice how arithmetical and/or manipulative errors can lead you astray.

If you have some of the other books of this series, try and make up mark schemes for the worked examples. You will find this process will assist you in writing out good solutions and understanding how you obtain credit for your efforts.

4 Multiple-choice tests: pure mathematics

4.1 Answering techniques

In multiple-choice papers a large number of questions have to be done in the allocated time. Hence, it is essential to work quickly and to recognize any possible short cuts. In this chapter the two main styles of multiple-choice question are considered in detail and the various methods of approach are described.

I Simple multiple-choice

In this type of question a problem is posed and five possible answers, **A, B, C, D** and **E**, are given. The candidate is required to identify the correct answer to the problem. [In multiple-choice questions the correct answer is known as the 'key'.]

Methods

Owing to the short time available, approximately two minutes per question, various methods of solution should be tried.

(i) Work the question in the ordinary way and find the correct answer from those supplied. This is usually the best method in numerical questions and in simple straightforward questions.

Example 1 Two cards are to be drawn at random and without replacement, from an ordinary pack of 52 playing cards. The probability that they will both be picture cards is

A 33/676   **B** 11/221   **C** 9/169

D 20/221   **E** 6/13

First card, $P(\text{picture}) = \dfrac{12}{52} = \dfrac{3}{13}$,

Second card, $P(\text{picture}) = \dfrac{11}{51}$,

Answer $= \dfrac{3}{13} \cdot \dfrac{11}{51} = \dfrac{11}{221}$.

Hence key **B**.

(ii) Trial and error. Try each of the answers to see which one 'fits'. This method is suitable where the actual working out of the problem would be long, or difficult, or both.

Example 2 The point $(5, 3)$ is one end of a diameter of the circle $(x - 2)^2 + (y + 1)^2 = 25$. The coordinates of the other end of this diameter are

A $(-1, 5)$ **B** $(1, -5)$ **C** $(-5, -3)$

D $(-1, -5)$ **E** $(5, 3)$

The centre of the circle is $(2, -1)$ and this must clearly be the mid-point of the ends of the diameter. Testing the mid-point of the join of $(5, 3)$ with each of the points in turn, we get $(2, 4)$ no good; $(3, -1)$ no good; $(0, 0)$ no good; $(2, -1)$ correct, giving key **D**. Do not waste time testing the y-coordinate if the x-coordinate does not apply. [This method can be a real time-saver when the key happens to be **A** or **B**.]

(iii) Elimination. Some of the suggested answers can often be discarded for a variety of reasons, leaving either many fewer answers to be considered or just one answer which is the correct one.

Example 3 An equation of the curve which passes through the point $(1, 1)$, and for which $\dfrac{dy}{dx} = 4x + 2$, is

A $y = 2x^2 + 2x$ **B** $y = x^2 + 2x - 2$

C $y = 2x^2 + 2x - 2$ **D** $y = 2x^2 + 2x - 3$

E $y = 2x^2 + 2x - 1$

The coordinates of the point $(1, 1)$ do not satisfy the equations in **A**, **C** or **E**, so this leaves only **B** and **D** to be considered and we can quickly see that only for **D** of these is $\dfrac{dy}{dx} = 4x + 2$. Hence the key is **D**.

Example 4 $\dfrac{x^3 + x^2 + x - 1}{(x^2 + 1)(x - 1)}$ can be expressed, with P, Q, R, M, N non-zero constants, in the form

A $\dfrac{P}{x^2 + 1} + \dfrac{Q}{x - 1}$ **B** $\dfrac{Px + R}{x^2 + 1} + \dfrac{Q}{x - 1}$

C $\dfrac{Px}{x^2+1}+\dfrac{Q}{x-1}$ **D** $Mx+N+\dfrac{Px}{x^2+1}+\dfrac{Q}{x-1}$

E $N+\dfrac{Px+R}{x^2+1}+\dfrac{Q}{x-1}$

Finding the partial fractions can be quite a lengthy business, but, if we observe that the degree of numerator and denominator is the same, i.e. cubic, then we known that we must first divide numerator by denominator. This immediately dismisses **A, B** and **C** in which no such division has been done and also **D** for which a power of x^4 would have been needed in the numerator. This leaves only **E** which must, therefore, be the key.

(iv) Identifying some important points (as in graphical questions, for instance) or sometimes a partial working of the problem will enable one to identify the correct key.

Example 5

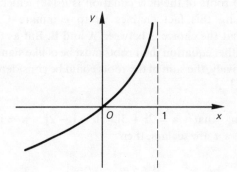

The graph could be a sketch of the curve

A $y=\dfrac{1}{1-x}$ **B** $y=\dfrac{1}{x-1}$ **C** $y=-\ln(1-x)$

D $y=\dfrac{x}{1-x}$ **E** $y=\dfrac{x}{x-1}$

The important points we note immediately from the given graph are:

(a) point $(0,0)$ lies on the curve,
(b) the line $x=1$ is an asymptote,
(c) there are no values of y for $x>1$.

Hence

(a) dismisses **A** and **B** as neither passes through $(0,0)$,
(b) satisfies **C, D** and **E**, so look at (c), which dismisses **D** and **E**, so the key is **C**.

[This is, of course, the process of elimination again but we do need to pick out the important points in order to use this method.]

Example 6 Given that the roots of the equation $ax^2 + bx + c = 0$, where $abc \neq 0$, are α and β, then the roots of the equation $4cx^2 + 2bx + a = 0$ are

A $\dfrac{1}{2\alpha}$ and $\dfrac{1}{2\beta}$ **B** $-\dfrac{1}{2\alpha}$ and $-\dfrac{1}{2\beta}$ **C** $\dfrac{\alpha}{2}$ and $\dfrac{\beta}{2}$

D $\dfrac{2}{\alpha}$ and $\dfrac{2}{\beta}$ **E** 2α and 2β

Most candidates would probably start by writing

$$\alpha + \beta = -\frac{b}{a}, \qquad \alpha\beta = \frac{c}{a}.$$

There is no need, however, to work the whole problem out if we note that the product of the roots of the new equation is $a/(4c)$ which is equal to $1/(4\alpha\beta)$. Recognizing this fact enables us to eliminate **C**, **D** and **E** immediately, so that the choice is between **A** and **B**. But as there are no negative signs in either equation their roots must be of like signs. Hence the key is **A**. [Alternatively, the sum of the roots could be considered to decide between **A** and **B**.]

Example 7 Given that $\mathbf{a} = 2\mathbf{i} + 3\mathbf{j}$, $\mathbf{b} = \mathbf{i} - 2\mathbf{j}$, $\mathbf{x} = \mathbf{i} + 5\mathbf{j}$ and $\mathbf{x} = s\mathbf{a} + t\mathbf{b}$ where s, t are scalars, then

A $s = 1, t = 1$ **B** $s = -1, t = 1$

C $s = -1, t = -1$ **D** $s = 1, t = -1$

E the values of s and t cannot be found

This could be done by the 'trial and error' process but it is probably better to go some way with the working first:

$$\mathbf{i} + 5\mathbf{j} = s(2\mathbf{i} + 3\mathbf{j}) + t(\mathbf{i} - 2\mathbf{j})$$
$$\Leftrightarrow \quad \mathbf{i} + 5\mathbf{j} = (2s + t)\mathbf{i} + (3s - 2t)\mathbf{j}$$
$$\Leftrightarrow \quad 2s + t = 1,$$
$$3s - 2t = 5.$$

These equations are, of course, easily solved but it is probably quicker to observe that **A**, **B** and **C** do not satisfy the first equation whereas **D** does and then we can seen at once that **D** also satisfies the second equation. Hence the key is **D**.

Note that in some cases key **E**, often written as 'none of the above', may be correct. It must not be automatically discarded.

Example 8 $\dfrac{\ln 40}{\ln 8} =$

A $\ln 5$ **B** $\ln 32$

C $\ln 40 - \ln 8$ **D** $1 + \ln\left(\dfrac{5}{8}\right)$

E   none of the above

$\dfrac{\ln 40}{\ln 8} = \dfrac{\ln 8 + \ln 5}{\ln 8} = 1 + \dfrac{\ln 5}{\ln 8}$ so that **E** is clearly the correct key.

II Multiple-completion

This type of question is not a single problem to be solved but consists of some given information and then three different relating assertions are made. The candidate is required to investigate each of these assertions to find whether or not it is correct. The choices of key are then as follows:

	Summary		
A	1	2	3
B	1	2	
C		2	3
D	1		
E			3

A   if **1, 2** and **3** are correct

B   if only **1** and **2** are correct

C   if only **2** and **3** are correct

D   if only **1** is correct

E   if only **3** is correct

This appears to mean that three separate problems are involved in these questions. However, the three statements made will usually have a common theme, for instance features of a graph or roots of an equation, and often a little work done on the data will give us all the answers. This is best illustrated by examples.

Example 9 $z = \dfrac{3 - i}{1 + 3i}.$

1 $|z| = 1$

2 $zz^* = 1$

3 $z + z^* = 0$

The work we need to do on the data here is to write z in the form $a + bi$, thus:

$$z = \frac{(3 - i)(1 - 3i)}{(1 + 3i)(1 - 3i)} = \frac{-10i}{10}$$

$$\Rightarrow \quad z = -i \quad \Rightarrow \quad z^* = i.$$

It is then easily seen that **1**, **2** and **3** are all correct so that the key is **A**.

Example 10 The curve $y = \dfrac{x}{x + 1}$ has

1 only one asymptote

2 no stationary points

3 no point of inflexion

It is a good rule when dealing with rational functions always to divide the numerator by the denominator if this is possible, i.e. if the numerator is of the same or higher degree than the denominator.

In this case $y = 1 - \dfrac{1}{x + 1}$.

When $\left. \begin{array}{l} x \rightarrow -1, y \rightarrow -\infty \\ \text{and when} \quad x \rightarrow \infty, \ y \rightarrow 1 \end{array} \right\} \Rightarrow x = -1$ and $y = 1$ are asymptotes

so **1** is incorrect.

For turning points,

$$\frac{\mathrm{d}y}{\mathrm{d}x} = (x + 1)^{-2}$$

$$= \frac{1}{(x + 1)^2} \neq 0 \qquad \text{for } x \in \mathbb{R}.$$

Hence no stationary points, so **2** is correct.

$$\frac{\mathrm{d}^2 y}{\mathrm{d}x^2} = -\frac{2}{(x + 1)^3} \neq 0 \qquad \text{for } x \in \mathbb{R}.$$

Hence no inflexions, so **3** is correct. The key is therefore **C**.

In this example the simple rewriting of y in the form suggested not only made the asymptotes more obvious but made the repeated differentiation considerably easier.

It is worth noticing that not all the logical possibilities are present in the keys. Thus **2** only or **1** and **3** only are not possible keys. This means that when **1** is incorrect then **3** must be correct.

Note: If **1** is correct, key cannot be **C** or **E**.

If **1** is wrong, **2** is wrong, key must be **E**; there is no need to check that **3** is correct.

If **1** is wrong, **2** is correct, key must be **C**; there is no need to check that **3** is correct.

A good diagram often supplies all that is needed to answer a question.

Example 11   The equation $e^{-x} = \sin x$, where $x \in \mathbb{R}$,

1   has only positive roots

2   has an infinite number of roots

3   has two roots in the interval $(0, \pi)$

For this type of equation, which has no easy method of solution, a graphical method provides a great deal of information about the roots. In this case the roots are shown graphically as the values of x at the intersection of the curves $y = e^{-x}$ and $y = \sin x$. Both of these are curves which candidates are expected to be able to draw readily.

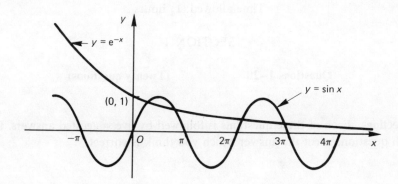

The diagram enables us to see that all three statements are true and so the key is **A**.

Example 12   An equation of a circle is

$$x^2 + y^2 - 2x + 2y + 1 = 0.$$

1 The circle touches both coordinate axes

2 For any point (x, y) on the circle, $x < 0$ and $y < 0$

3 The centre of the circle lies on the line $y = x$

Expressing the equation in the form

$$(x - 1)^2 + (y + 1)^2 = 1,$$

we see that the centre is $(1, -1)$ and the radius is 1 so that the diagram is

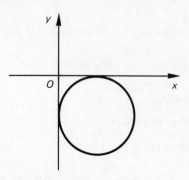

The key is then **D** as only **1** is correct.

4.2 Paper 1 (multiple-choice)

Time allowed: $1\frac{1}{4}$ hours

SECTION I

Questions 1–20   **(Twenty questions)**

Directions. Each of these questions is followed by five suggested answers. For each question, select the answer which you think is correct.

1 $\cos 3\theta - \cos 7\theta \equiv$

     A $-2 \sin 2\theta \sin 5\theta$ **B** $2 \sin 2\theta \sin 5\theta$

     C $2 \cos 2\theta \cos 5\theta$ **D** $2 \sin 2\theta \cos 5\theta$

     E $-2 \sin 2\theta \cos 5\theta$

2 A circle, whose centre is the point $(a, -b)$, passes through the origin. An equation of the circle is

A $x^2 + y^2 + ax - by = 0$ **B** $x^2 + y^2 - ax + by = 0$

C $x^2 + y^2 + 2ax - 2by = 0$ **D** $x^2 + y^2 - 2ax + 2by = 0$

E $x^2 + y^2 - 2ax + 2by = a^2 + b^2$

3 Given that $\mathbf{a} = (-3\mathbf{i} + 2\mathbf{j} + 5\mathbf{k})$, $\mathbf{b} = (5\mathbf{i} + 3\mathbf{j} - 2\mathbf{k})$, then $\mathbf{a} \cdot \mathbf{b} =$

A 38 **B** 27 **C** 19 **D** -19 **E** -27

4 Given that $|z| = 2$ and $\arg z = -2\pi/3$, then $z =$

A $-1 - i\sqrt{3}$ **B** $-1 + i\sqrt{3}$ **C** $-\sqrt{3} - i$

D $-\sqrt{3} + i$ **E** $-2 - 2i\sqrt{3}$

5 $\displaystyle\int \frac{1}{9 + 16x^2}\,dx =$

A $\frac{1}{12}\tan^{-1}\left(\dfrac{4x}{3}\right) + \text{constant}$ **B** $\frac{1}{4}\tan^{-1}\left(\dfrac{4x}{3}\right) + \text{constant}$

C $\frac{4}{3}\tan^{-1}\left(\dfrac{4x}{3}\right) + \text{constant}$ **D** $\frac{1}{4}\sin^{-1}\left(\dfrac{4x}{3}\right) + \text{constant}$

E $\frac{1}{12}\sin^{-1}\left(\dfrac{4x}{3}\right) + \text{constant}$

6 The number of permutations of the six letters of the word PLASMA is

A 90 **B** 180 **C** 360

D 720 **E** none of the above

7 $\displaystyle\sum_{r=1}^{\infty} (-1)^r \left(\frac{4}{5}\right)^r =$

A 5 **B** $\frac{5}{9}$ **C** $\frac{4}{9}$ **D** $-\frac{4}{9}$ **E** -4

8 $\cos \theta° = \cos 310°$ implies that, for $n \in \mathbb{Z}$, $\theta =$

A $180n - 50$ **B** $180n + (-1)^n 50$ **C** $180n \pm 50$

D $360n - 50$ **E** $360n \pm 50$

9 Given that $x = t^3$, $y = t^2$, then $\dfrac{d^2 y}{dx^2} =$

A $\dfrac{2}{3t}$ **B** $-\dfrac{2}{9t^4}$ **C** $-\dfrac{2}{3t^2}$ **D** $\dfrac{4}{9t^2}$ **E** $\dfrac{1}{3t}$

10 $\dfrac{d}{dx} \ln(4x^3) =$

A $\dfrac{3}{x}$ **B** $\dfrac{1}{4x^3}$ **C** $\dfrac{1}{12x^2}$ **D** $\dfrac{12}{x}$ **E** $\dfrac{4}{3x^2}$

11 The remainder when $(x^3 + 2x^2 - x + 6)$ is divided by $(x + 3)$ is

A -6 **B** 0 **C** 48 **D** 54

E none of the above

12 Given that $y^2 = 4x^3$, which one of the following does not give a straight line graph?

A y^2/x^2 plotted against x **B** $\log y$ plotted against $\log x$

C $\dfrac{1}{y^2}$ plotted against $\dfrac{1}{x^3}$ **D** xy^2 plotted against x^4

E $\log x$ plotted against y^2

13

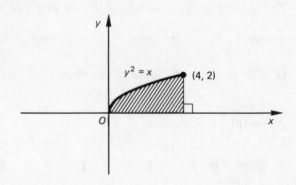

The area, in units^2, of the shaded region is

A $\dfrac{2\sqrt{2}}{3}$ **B** $\dfrac{4\sqrt{2}}{3}$ **C** $\frac{8}{3}$ **D** $\frac{16}{3}$ **E** 8

14 The complete set of values of x for which $6x^2 > 11x + 10$ is

A $\{x : -\frac{2}{3} < x < \frac{5}{2}\}$ **B** $\{x : x < -\frac{2}{3}\} \cup \{x : x > \frac{5}{2}\}$

C $\{x : -\frac{5}{2} < x < \frac{2}{3}\}$ **D** $\{x : x < -\frac{5}{2}\} \cup \{x : x > \frac{2}{3}\}$

E none of the above

15 The roots of the quadratic equation $ax^2 + bx + c = 0$, where $ac \neq 0$, are α and β.

$$\dfrac{1}{\alpha\beta^2} + \dfrac{1}{\alpha^2\beta} =$$

A $-\dfrac{b}{c}$ **B** $\dfrac{b}{c}$ **C** $-\dfrac{ab}{c^2}$ **D** $\dfrac{ab}{c^2}$

E none of the above

16 $\displaystyle\int_1^e \ln x \, \mathrm{d}x =$

A -1 **B** e **C** $e - 1$ **D** $e + 1$ **E** 1

17 The gradient of the normal to the curve $x^2 y = 1$ at the point $(1/t, t^2)$ is

A $-2t^3$ **B** $-\dfrac{1}{2t^3}$ **C** $\dfrac{1}{2t^3}$ **D** $2t^3$

E none of the above

18

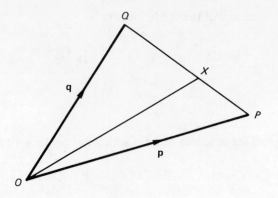

P and Q have positive vectors $\mathbf{p}, \mathbf{q}$ respectively relative to the origin O. Given that X is the point in PQ such that $PX:XQ = 2:3$, then $\overrightarrow{OX} =$

A $\frac{1}{5}(3\mathbf{p} + 2\mathbf{q})$ **B** $\frac{1}{5}(2\mathbf{p} + 3\mathbf{q})$ **C** $\frac{1}{3}(\mathbf{p} + 2\mathbf{q})$

D $\frac{1}{3}(2\mathbf{p} + \mathbf{q})$ **E** $\frac{1}{5}(7\mathbf{p} + 2\mathbf{q})$

19 Given that α is a first approximation to a root of the equation $f(x) = 0$, then, in general, a better approximation given by one application of the Newton–Raphson method is

A $[\alpha f'(\alpha) + f(\alpha)]/f'(\alpha)$ **B** $[\alpha f(\alpha) - f'(\alpha)]/f(\alpha)$

C $[\alpha f'(\alpha) - f(\alpha)]/f(\alpha)$ **D** $[f'(\alpha) - \alpha f(\alpha)]/f'(\alpha)$

E none of the above

20 The graph of $y = x\,e^{-x}$ could be

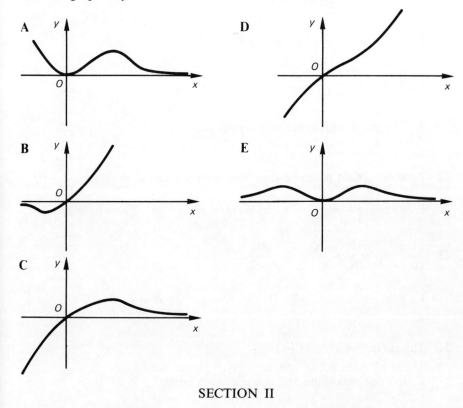

A

B

C

D

E

Questions 21–30   **(Ten questions)**

Directions. For each of the following questions, **ONE** or **MORE** of the responses given are correct. Decide which of the responses is (are) correct. Then choose

A   if **1, 2** and **3** are correct

B   if only **1** and **2** are correct

C   if only **2** and **3** are correct

D   if only **1** is correct

E   if only **3** is correct

Directions summarized			
A	1	2	3
B	1	2	
C		2	3
D	1		
E			3

21 $f : x \mapsto \dfrac{x}{x-1}, \quad x \in \mathbb{R}, x \neq 1.$

   1 $f^{-1} : x \mapsto \dfrac{x-1}{x}, \quad x \in \mathbb{R}, x \neq 0$

   2 $ff : x \mapsto \left(\dfrac{x}{x-1}\right)^2, \quad x \in \mathbb{R}, x \neq 1$

   3 The graph of $f(x)$ has two asymptotes

22 $7 \cos \phi + 24 \sin \phi$ can be expressed, with α acute, in the form

   1 $25 \cos (\phi - \alpha)$ **2** $25 \sin (\phi + \alpha)$ **3** $25 \cos (\phi + \alpha)$

23 $\dfrac{1}{x(x^2 - 4)} \equiv \dfrac{P}{x} + \dfrac{Q}{x+2} + \dfrac{R}{x-2}.$

   1 $P = -\frac{1}{4}$ **2** $Q = \frac{1}{8}$ **3** $R = \frac{1}{8}$

24 The curve $y = x^3 - 3x - 3$

   1 has one maximum and one minimum point

   2 cuts the x-axis in three points

   3 has no points of inflexion

25 **1** $\displaystyle\int_0^{\pi/4} \sin^2 x \, dx = \int_0^{\pi/4} \cos^2 x \, dx$

   2 $\displaystyle\int_0^{\pi/2} \sin^3 x \, dx = \int_0^{\pi/2} \cos^3 x \, dx$

   3 $\displaystyle\int_0^{\pi} \sin^4 x \, dx = \int_0^{\pi} \cos^4 x \, dx$

26 When expanded in ascending powers of x,

$$(8 - 4x)^{-1/3} = a_0 + a_1 x + a_2 x^2 + \ldots .$$

   1 $a_0 = \frac{1}{2}$ **2** $a_1 = -\frac{1}{12}$ **3** $a_2 = \frac{1}{12}$

27 For all complex numbers z_1, z_2

    1 $|z|^2 = zz^*$ **2** $|z_1 + z_2| \leqslant |z_1| + |z_2|$

    3 $\arg(z_1 z_2) = (\arg z_1)(\arg z_2)$

28 The curves $y = e^x$ and $y = \ln x$

    1   do not intersect

    2   do not have stationary points

    3   are mirror images of each other in the line $y = x$

29 The number of different solutions of the equation $3 \sin \theta + 4 \cos \theta = k$ in the range $0 \leqslant \theta < 2\pi$ is

    1   two if $|k| > 5$ **2**   zero if $|k| < 5$

    3   one if $|k| = 5$

30 Given that $\mathbf{a}, \mathbf{b}, \mathbf{c}$ are constant non-zero and non-parallel vectors, and $\mathbf{r}$ is the position vector of a variable point P, which of the following equations imply that P lies in a fixed plane?

    1 $\mathbf{r} = \mathbf{a} + t(\mathbf{c} - \mathbf{b})$,   where t is a parameter

    2 $\mathbf{r} = \mathbf{a} + s\mathbf{b} + t\mathbf{c}$,   where s and t are parameters

    3 $(\mathbf{r} - \mathbf{a}) \cdot (\mathbf{b} + \mathbf{c}) = 0$

4.3   Solutions for Paper 1

    1   Simple recall of formula; note that you are probably accustomed to

$$\cos 7\theta - \cos 3\theta \equiv -2 \sin 2\theta \sin 5\theta,$$

but this has the opposite sign, i.e.

$$\cos 3\theta - \cos 7\theta \equiv 2 \sin 2\theta \sin 5\theta.$$

2 Care with signs needed here. Use the general equation of the circle

$$x^2 + y^2 + 2gx + 2fy + c = 0,$$

where $(-g, -f)$ is the centre.
Hence $g = -a$, $f = b$, and $c = 0$ [as $(0, 0)$ lies on the circle]. Hence key **D**.
Alternatively, use $(x - a)^2 + (y + b)^2 = a^2 + b^2$.

3 $\mathbf{a} . \mathbf{b} = -15 + 6 - 10 = -19$, i.e. the **i** with the **i**, the **j** with the **j** and the **k** with the **k**.

4 A diagram gives the shortest solution here and is probably the least susceptible to error:

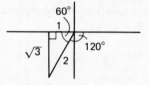

If preferred,

$$z = 2\left[\cos\left(\frac{-2\pi}{3}\right) + i \sin\left(\frac{-2\pi}{3}\right) \right] = 2(-\tfrac{1}{2} - i\sqrt{3}/2)$$
$$= -1 - i\sqrt{3}.$$

However, the diagram is safer for most candidates.

5 $\int \dfrac{1}{9 + 16x^2} \, dx$. This is a standard integral which should be recognized. However, most people are aware that the answer is 'tan^{-1}' but are perhaps a little vague on the constants. Try

$$\frac{d}{dx} \tan^{-1}\left(\frac{4x}{3}\right) = \frac{1}{1 + (16x^2)/9} \cdot \frac{4}{3} = \frac{1}{9 + 16x^2} \cdot 12.$$

So the required integral is $\frac{1}{12} \tan^{-1}\left(\dfrac{4x}{3}\right) +$ constant.

6 This should be a known formula: $6!/2! = 360$.

7 $\displaystyle\sum_{r=1}^{\infty} (-1)^r (\tfrac{4}{5})^r = -\tfrac{4}{5} + (\tfrac{4}{5})^2 - (\tfrac{4}{5})^3 + \dots$

$$= (-\tfrac{4}{5}) + (-\tfrac{4}{5})^2 + (-\tfrac{4}{5})^3 + \dots .$$

If you do not recognize the series immediately, the first two or three terms are a help to see that it is a geometrical progression.

$$S_\infty = \frac{-\tfrac{4}{5}}{1 - (-\tfrac{4}{5})} = -\tfrac{4}{9}.$$

8 $\cos \theta° = \cos 310°$. The general solution is required. Notice that all the

answers are in terms of $50°$ and we see that

$\cos 310° = \cos(-50°) = \cos 50°$ so

$\cos \theta° = \cos 50°$

$\Leftrightarrow \ \theta = 360n \pm 50$ (the general solution should be known).

9 Care is needed with this one. Use

$$\frac{dy}{dx} = \frac{dy}{dt} \bigg/ \frac{dx}{dt} \quad \Rightarrow \quad \frac{dy}{dx} = \frac{2t}{3t^2} = \frac{2}{3t}.$$

Then

$$\frac{d^2y}{dx^2} = \frac{d}{dt}\left(\frac{dy}{dx}\right) \bigg/ \frac{dx}{dt} = \left(-\frac{2}{3t^2}\right) \bigg/ (3t^2) = -\frac{2}{9t^4}.$$

$$\left[\text{What you must } not \text{ say is } \frac{d^2y}{dx^2} = \frac{d^2y}{dt^2} \bigg/ \frac{d^2x}{dt^2}. \right]$$

10 $\dfrac{d}{dx} \ln(4x^3) = 12x^2 \cdot \dfrac{1}{4x^3} = \dfrac{3}{x}$ or $\dfrac{d}{dx}\left[\ln 4 + 3 \ln x\right] = \dfrac{3}{x}.$

11 Remainder $= f(-3) = -27 + 18 + 3 + 6 = 0.$

12 The more experienced candidate will probably 'spot' **E** at once because of the impossibility of getting log x without log y from the given equation. It is always worth glancing through the answers to see if one is blatantly impossible as in this case. However, it does not take long to rearrange (preferably mentally) the equation to give a linear one in each case, thus:

$$y^2/x^2 = 4x,$$

$$2 \log y = \log 4 + 3 \log x,$$

$$1/x^3 = 4/y^2,$$

$$xy^2 = 4x^4.$$

13 $\text{Area} = \displaystyle\int_0^4 x^{1/2}\, dx = \left[\frac{2}{3}x^{3/2}\right]_0^4 = \frac{16}{3}.$

14 In most inequality questions (involving rational functions) collect all the terms on one side as it is usually easier to test the sign of an expression rather than finding the values of x for which one expression is greater than another. (An exception to this occurs when a graphical method is clearly called for.)

$$6x^2 - 11x - 10 > 0.$$

In most such examples the quadratic is factorized, but if factors are not obvious make sure whether there are real factors by testing the sign of '$b^2 - 4ac$'. In this case $121 + 240 > 0$, so factors possible and are $(3x + 2)(2x - 5)$. [The appearance of $\frac{2}{3}, \frac{5}{2}$ etc. in the solutions is a guide.] Then solution

$$\{x : x < -\tfrac{2}{3}\} \cup \{x : x > \tfrac{5}{2}\}.$$

15 $\alpha + \beta = -b/a, \alpha\beta = c/a.$ $\qquad \dfrac{\alpha + \beta}{\alpha^2\beta^2} = -\dfrac{b}{a} \cdot \dfrac{a^2}{c^2} = -\dfrac{ba}{c^2}.$

16 $\displaystyle\int_1^e \ln x\, dx = \left[x \ln x\right]_1^e - \int_1^e x \cdot \frac{1}{x}\, dx \qquad \text{(by parts)}$

$$= e - e + 1 = 1.$$

17 $y = \dfrac{1}{x^2} \ \Rightarrow\ \dfrac{dy}{dx} = -2/x^3$

$$\Rightarrow\ \text{Gradient of normal} = (-1)\Big/\frac{dy}{dx} = \frac{x^3}{2} = \frac{1}{2t^3}.$$

Alternatively,

$$\frac{dx}{dt} = -\frac{1}{t^2}, \ \frac{dy}{dt} = 2t \ \Rightarrow\ \frac{dy}{dx} = -2t^3 \quad \text{etc.}$$

18 The formula here, $\dfrac{(k\mathbf{b} + l\mathbf{a})}{k + l}$, should be known.

$$\overrightarrow{OX} = (2\mathbf{q} + 3\mathbf{p})/5.$$

19 Simple recall of Newton–Raphson formula gives

$$\alpha - \frac{f(\alpha)}{f'(\alpha)},$$

which is not given, hence key is **E**.

20 Look out for special features of the curve. Passes through O—all sketches do.

$y/x = e^{-x} > 0$, so $x > 0$, $y > 0$ or $x < 0$, $y < 0$, and this cuts out **A** and **E**. When $x \to \infty$, $y \to 0$, i.e. $y = 0$ is an asymptote. This cuts out **B** and **D**. Hence the key is **C**.

We note that the turning point(s) are also important features, but it would take time to find the stationary value(s), so we try to manage without doing this.

21 $f : x \mapsto \dfrac{x}{x - 1}$. For f^{-1}, say $y = \dfrac{x}{x - 1} \Rightarrow xy - y = x$

$$\Rightarrow x = \frac{y}{y - 1},$$

i.e. function is self-inverse and **1** is incorrect.

$$ff : x \mapsto \frac{\dfrac{x}{x - 1}}{\dfrac{x}{x - 1} - 1} = \frac{x}{\not{x} - \not{x} + 1} \quad \text{and } \mathbf{2} \text{ is incorrect.}$$

It follows that **3** must be correct and the key is **E** but, it is useful to check that there are two asymptotes, $x = 1$ and $y = 1$. (Do not forget the possibility of an asymptote parallel to the x-axis, $y = 1 + \dfrac{1}{x - 1}$ and $y \to 1$ when $x \to \infty$.)

22 $25 \cos (\phi - \alpha) = 25 \cos \phi \cos \alpha + 25 \sin \phi \sin \alpha$

$$= 7 \cos \phi + 24 \sin \phi, \quad \text{i.e. } \mathbf{1} \text{ is correct.}$$

$25 \sin (\phi + \alpha) = 25 \sin \phi \cos \alpha + 25 \cos \phi \sin \alpha$

$$= 24 \sin \phi + 7 \cos \phi, \quad \text{i.e. } \mathbf{2} \text{ is correct.}$$

There is no point in trying **3** as, given that α is acute, the signs will clearly be wrong. You should, in fact, be able to see without working that **2** must be correct if **1** is correct.

23 Try to practise the 'cover-up' method for doing simple partial fractions quickly:

$$\frac{1}{x(x + 2)(x - 2)}.$$

For P, cover x with your finger and put $x = 0$, giving $P = -\frac{1}{4}$.
For Q, cover $(x + 2)$ with your finger and put $x = -2$, giving $Q = \frac{1}{8}$.
For R, cover $(x - 2)$ with your finger and put $x = 2$, giving $R = \frac{1}{8}$. Key **A**.

24 $y = x^3 - 3x - 3$.

$\dfrac{\mathrm{d}y}{\mathrm{d}x} = 3x^2 - 3 = 0$ for stationary values, giving $x = \pm 1$ and so $y = -5, -1$.

Students should be aware of the form of the cubic curve and the fact that, when there are two stationary values, one must be a maximum and one a minimum, without further working. Also, as the curve is continuous, if the maximum and minimum values of y have the same sign the curve cannot cut the x-axis in three points. Further, when there is a maximum and a minimum value in a continuous curve there must be a point of inflexion between them where the curve changes direction. Hence the key is **D**. There is no real need to draw a sketch to see these points, and so the only working needed is that given in the first three lines above. (Watch out for the 'double negative' which often causes confusion. There *is* a point of inflexion so that it is *incorrect* to say that there are *no* points of inflexion.)

25 In each case put $x = \pi/2 - z$.

1 $\text{LHS} = -\displaystyle\int_{\pi/2}^{\pi/4} \cos^2 z \, \mathrm{d}z = \int_{\pi/4}^{\pi/2} \cos^2 z \, \mathrm{d}z \neq \text{RHS.}$

2 $\text{LHS} = -\int_{\pi/2}^{0} \cos^3 z \, dz = \int_{0}^{\pi/2} \cos^3 z \, dz = \text{RHS}.$

It follows that **3** is correct as **2** alone is not a key, but to check:

$$\text{LHS} = -\int_{\pi/2}^{-\pi/2} \cos^4 x \, dx = \int_{-\pi/2}^{\pi/2} \cos^4 x \, dx$$

$$= 2\int_{0}^{\pi/2} \cos^4 x \, dx \quad \text{(even function)}$$

$$= \int_{0}^{\pi} \cos^4 x \, dx \quad \text{(symmetry about } x = \pi/2\text{)}.$$

Key **C**.

26 $(8 - 4x)^{-1/3} = 8^{-1/3}(1 - \tfrac{1}{2}x)^{-1/3}$

$$= \tfrac{1}{2}(1 - \tfrac{1}{2}x)^{-1/3} = \frac{1}{2}\left[1 + \tfrac{1}{6}x + \frac{(-\tfrac{1}{3})(-\tfrac{4}{3})}{1 \cdot 2} \cdot \tfrac{1}{4}x^2 + \ldots\right].$$

Hence **1** is correct, **2** incorrect, **3** incorrect, so the key is **D**.

(There is no need to work out the coefficient of x^2—just observe that it is clearly not $\tfrac{1}{12}$.)

27 This is a collection of facts about complex numbers which students should know.

1 Conjugates differ only in sign and so their moduli are equal.

2 The length of one side of a triangle $\leqslant$ sum of the lengths of the other two and so this is true.

3 Incorrect, arg(product) = *sum* of arguments, *not* product of arguments.

Key **B**.

28 These are well-known inverse curves and should be quickly sketched:

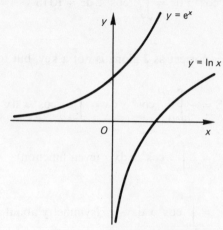

This gives key **A**. (**3** does, in fact, follow from the fact that the functions are inverse.)

29 Put the equation in the form

$$5 \sin (\theta + \alpha) = k, \quad \text{where } \cos \alpha = \tfrac{3}{5}, \sin \alpha = \tfrac{4}{5}$$

(although the values of α are not needed). The question is based on the fact that $|\sin x| \leqslant 1$, giving **1** and **2** both incorrect, and so **3** is correct, giving key **E**.

30 The equation of a plane given in parametric form must have *two* parameters, so **1** is incorrect while **2** is correct. **3** is also correct as it is a correct perpendicular form of the equation of a plane. There is no need to identify the vectors, simply to see whether the equations are feasible as the equations of a plane. Key **C**.

4.4 Paper 2 (multiple-choice)

Time allowed: $1\tfrac{1}{4}$ hours

SECTION I

Questions 1–20   **(Twenty questions)**

Directions. Each of these questions is followed by five suggested answers. For each question select the answer which you think is correct.

1 $\sin 3\theta - \sin 5\theta \equiv$

A $2 \sin 4\theta \cos \theta$ **B** $2 \sin \theta \sin 4\theta$ **C** $2 \cos 4\theta \sin \theta$

D $-2 \cos 4\theta \sin \theta$ **E** $-2 \sin 4\theta \cos \theta$

2 $\overrightarrow{OP} = 4\mathbf{i} + 3\mathbf{j},\ \overrightarrow{OQ} = 3\mathbf{i} + 4\mathbf{j}.$
$|\overrightarrow{PQ}| =$

A $7\sqrt{2}$ **B** $\sqrt{2}$ **C** 2 **D** 25 **E** 24

3 The number of combinations of five different things taken three at a time is

A 10 **B** 15 **C** 20 **D** 60 **E** 120

4 Given that $x^y = y^z$, where $x, y, z \in \mathbb{R}^+$, then

A $\log_x y = \log_y z$ **B** $y \ln z = x \ln y$

C $\log_y x = \log_z y$ **D** $z \log_y x = y$

E $y \ln x = z \ln y$

5 The roots of the equation $x^2 + 5x + 4 = 0$ are α, β.
$\alpha^2 + \beta^2 =$

A 33 **B** 21 **C** 17 **D** 14 **E** -17

6 The circles C_1 and C_2 have equations $x^2 + y^2 - 4y = 0$ and $x^2 + y^2 - 4x = 0$ respectively. The point $(1, 1)$ lies

A outside both circles   **B** inside both circles

C on the circumference of C_1 **D** inside C_1 but outside C_2

E inside C_2 but outside C_1

7 Given that $y = (\ln x)^3$, then

$$\frac{dy}{dx} =$$

A $\dfrac{3 \ln x}{x}$ B $\dfrac{3}{x^2}$ C $3 \ln x$

D $3(\ln x)^2$ E $\dfrac{3(\ln x)^2}{x}$

8 Given that $|z| = 4$ and $\arg z = -2\pi/3$, then

$z =$

A $-2 - i2\sqrt{3}$ B $-2 + i2\sqrt{3}$ C $-2\sqrt{3} - i2$

D $-2\sqrt{3} + i2$ E $-2 + i\sqrt{3}$

9 The complete solution set of the inequality
$$2x^2 - 3x - 5 < 0$$
is

A $\{x : -5/2 < x < 1\}$ B $\{x : x < -1\} \cup \{x : x > 5/2\}$

C $\{x : -1 < x < 5/2\}$ D $\{x : x < 5/2\} \cup \{x : x > 1\}$

E $\{x : 1 < x < 5/2\}$

10 All solutions of the equation $\sin \theta = \sin 2\alpha$ are obtained by taking all integer values of n in

$\theta =$

A $n\pi + (-1)^n \alpha$ B $2n\pi + (-1)^n 2\alpha$

C $n\pi + (-1)^n 2\alpha$ D $n\pi \pm 2\alpha$

E $2n\pi \pm 2\alpha$

11 When $\sin x = 3/5$, $\cos x$

A $= \frac{4}{5}$ **B** $= -\frac{4}{5}$ **C** $= \frac{4}{3}$ **D** $= -\frac{4}{3}$

E cannot be found; there is insufficient data to find $\cos x$

12 Given that $x = 1 + \sin \theta$ and $y = 1 - \cos \theta$ then
$$\frac{dy}{dx} =$$

A $-\tan \theta$ **B** $\cot \theta$ **C** $-\cot \theta$

D $\tan \theta$ **E** $\dfrac{\sin \theta + 1}{\cos \theta - 1}$

13 $\displaystyle\int \frac{1}{\sqrt{(9 - 4x^2)}} \, dx =$

A $\frac{1}{4} \sin^{-1} \left(\dfrac{2x}{3}\right) + $ constant **B** $\frac{1}{2} \sin^{-1} \left(\dfrac{2x}{3}\right) + $ constant

C $\frac{1}{3} \sin^{-1} \left(\dfrac{2x}{3}\right) + $ constant **D** $2 \sin^{-1} \left(\dfrac{2x}{3}\right) + $ constant

E $\frac{3}{2} \sin^{-1} \left(\dfrac{2x}{3}\right) + $ constant

14 An equation of the tangent to the curve $y^2 = x$ at the point $(4, 2)$ is

A $2x - y - 6 = 0$ **B** $x - 4y - 4 = 0$

C $x - y - 2 = 0$ **D** $x - 2y + 6 = 0$

E $x - 4y + 4 = 0$

15 The coefficient of x^2 in the expansion of $(1 - 3x)^{1/3}$, in ascending powers of x for $|x| < \frac{1}{3}$, is

A 1 **B** -1 **C** $\frac{1}{3}$ **D** $-\frac{2}{9}$ **E** 2

16 f(x) is an odd function such that

$$f(x) \equiv -x \qquad \text{for } 0 \leqslant x < 1,$$

$$f(x) \equiv x - 2 \qquad \text{for } 1 \leqslant x \leqslant 2.$$

A sketch of $y = f(x)$ for $-2 \leqslant x \leqslant 2$ could be

A

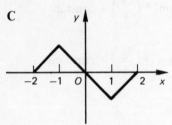

D

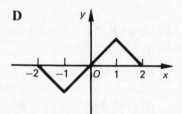

B

E

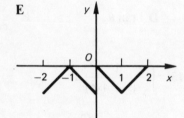

C

17

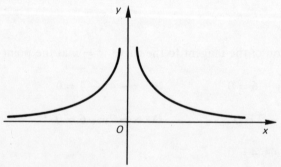

The graph could be a sketch of

A $xy = 1$ **B** $x^2y = 1$ **C** $xy^2 = 1$

D $y = 1 + 1/x^2$ **E** $y = 1 + 1/x$

18 If $x^2y^2 = 1 + 2x^2$, points on a straight line will be obtained by plotting

    A xy against x **B** $\dfrac{1}{x^2}$ against $\dfrac{1}{y^2}$ **C** xy against x^2

    D $\dfrac{1}{x^2}$ against y^2 **E**   none of the above

19 $\displaystyle\sum_{r=1}^{\infty} \left(\dfrac{-3}{4}\right)^r =$

    A $-\frac{3}{7}$ **B** $\frac{3}{7}$ **C** 3 **D** -3 **E** $-\frac{3}{4}$

20 $\displaystyle\int_{0}^{\pi/2} \sin x \cos x \, dx =$

    A $\frac{1}{4}$ **B** $\frac{1}{2}$ **C** 1 **D** 0 **E** -1

SECTION II

Questions 21–30   (Ten questions)

Directions. For each of the following questions, **ONE** or **MORE** of the responses given are correct. Decide which of the responses is (are) correct. Then choose

	Directions summarized		
A if **1, 2** and **3** are correct	**A**	1	2	3
B if only **1** and **2** are correct	**B**	1	2	
C if only **2** and **3** are correct	**C**		2	3
D if only **1** is correct	**D**	1		
E if only **3** is correct	**E**			3

21 $z = 3 + 4\text{i}$.

    1 $z - z^* = -8\text{i}$ **2** $z^* = \dfrac{1}{z}$ **3** $z^2 = -7 + 24\text{i}$

22 $\dfrac{1}{3 + 5x}$ can be expanded as a series of ascending powers of x when

    1 $x = -1$ **2** $x = \frac{1}{2}$ **3** $x = -\frac{1}{2}$

23

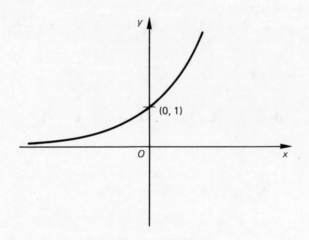

The graph could be a sketch of

    1 $y = \ln x$ **2** $y = \text{e}^x$ **3** $\log_a y = x$, where $a > 1$

24 The equation

$$\text{e}^x = \cos x$$

    1 has just one zero root   **2** has an infinite number of roots

    3 has no positive roots

25 The straight line

$$\mathbf{r} = \mathbf{i} - 2\mathbf{j} + 3\mathbf{k} + \lambda(2\mathbf{i} + \mathbf{j} + 2\mathbf{k}),$$

where λ is a parameter,

1 passes through the point with position vector $(-3\mathbf{i} - 4\mathbf{j} - \mathbf{k})$

2 is parallel to the vector $(-4\mathbf{i} - 2\mathbf{j} - 4\mathbf{k})$

3 is perpendicular to the vector $(3\mathbf{i} - 2\mathbf{j} - 2\mathbf{k})$

26 Given that

$$\frac{5x^2 - 4x + 5}{(x - 1)(x^2 + 2)} \equiv \frac{P}{x - 1} + \frac{Qx + R}{x^2 + 2},$$

where P, Q and R are constants, then

1 $P = 2$ **2** $Q = 3$ **3** $R = 1$

27 The parametric equations of a curve are $x = t^3$, $y = t^2$.

1 A cartesian equation of the curve is $x^2 = y^3$

2 The curve is symmetrical about the x-axis

3 The curve has zero gradient at the point $(0, 0)$

28

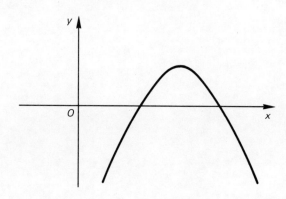

The diagram shows part of the curve $y = px^2 + qx + r$, where p, q and r are constants.

1 $p < 0$ **2** $r > 0$ **3** $q^2 < 4pr$

29 The angles of a triangle are x, y and z. Which of the following statements is (are) always true?

    1 $(x > y) \Rightarrow (\cos x < \cos y)$ **2** $(\sin x = \sin y) \Rightarrow (x = y)$

    3 $(x < y) \Rightarrow (\sin x < \sin y)$

30 $\displaystyle\int_a^b f(x)\,dx = P, \quad \int_a^b g(x)\,dx = Q,$ where $PQ \neq 0$ and $a < b$.

    1 $\displaystyle\int_a^b \left[f(x) + g(x)\right]\,dx = P + Q$

    2 $\displaystyle\int_a^b f(x)g(x)\,dx = PQ$

    3 $\displaystyle\int_a^b \left([f(x)]^2 + [g(x)]^2\right)\,dx = P^2 + Q^2$

5 Multiple-choice tests: applied mathematics

5.1 Strategy

In multiple-choice papers a large number of questions have to be done in the allocated time. Hence, it is essential to work quickly and to recognize any possible short cuts. In this chapter the two main styles of multiple-choice question are considered in detail and the various methods of approach described.

Simple multiple-choice

In this type of question a problem is posed and five possible answers, **A, B, C, D** and **E**, are given. The candidate is required to identify the correct answer to the problem. [In multiple-choice questions the correct answer is known as the 'key'.]

Methods

Owing to the short time available, approximately two minutes per question, various methods of solution should be tried.

(1) Work the question in the ordinary way and find the correct answer from those supplied. This is usually the best method in numerical questions and in simple straightforward questions.

Example 1 The kinetic energy, in J, of a body of mass 0·1 kg moving at speed $5 \, \text{m s}^{-1}$ is

A 0·125 **B** 0·25 **C** 0·5 **D** 1·25 **E** 2·5

The kinetic energy is $\frac{1}{2}mv^2$ giving

$$\tfrac{1}{2} \, . \, 0 \cdot 1 \times 5^2 \, \text{J} = 1 \cdot 25 \, \text{J}.$$

Hence key **D**.

(2) Straightforward recall. Sometimes a question merely requires you to remember a standard result. You should be able to 'spot' the correct answer at once.

Example 2 A particle, of mass m, moves in a straight line from the origin to a point P, position vector **p**, under the action of a constant force **F**. The work done by the force is

A $mp \cdot \mathbf{F}$ **B** $\mathbf{p} \cdot \mathbf{F}$ **C** $m|\mathbf{F}||\mathbf{p}|$

D $|\mathbf{F}||\mathbf{p}|$ **E** $|\mathbf{F}|\mathbf{p}$

The key is, of course, **B**. Note that the introduction of m is a 'red herring' and irrelevant to the question. Also, if in doubt about the answer, you should realize that its dimensions must be those of 'force' × 'distance' and so **A** and **C** can be discarded at once. The idea of dimensions must always be borne in mind when quantities such as m, $\mathbf{v}$, $\mathbf{F}$ containing units are used so that the actual units being used in the question are not specified.

(3) Trial and error. Try each of the answers to see which one 'fits'. This method is suitable where the actual working out of the problem would be long, or difficult, or both.

Example 3

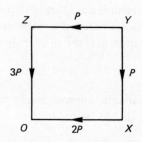

Four non-zero forces act as shown round the sides of a square $OXYZ$. The system

A is in equilibrium

B is equivalent to a force acting through Z

C is equivalent to a force acting through O

D is equivalent to a force acting through X

E is equivalent to a couple

We see that **A** is incorrect as resolving parallel to OX does not give zero, and **B** is incorrect as taking moments about Z does not give zero. For **C**, taking moments about O does give zero and we have already seen that the system is not in equilibrium and so the key is **C**.

(4) Elimination. Some of the suggested answers can often be discarded for a variety of reasons, leaving either many fewer answers to be considered or just one answer which is the correct one.

Example 4 A particle of mass m is moving vertically upwards under gravity in a medium in which the resistance is k times the speed. When the upward speed is v, the upward accleration is

A $-g - kv$ **B** $g - kv$ **C** $-g - kv/m$

D $g - kv/m$ **E** $g + kv/m$

The effect of gravity is to give a *downward* acceleration of magnitude g, i.e. an *upward* acceleration $-g$ so that we can discard keys **B**, **D** and **E**. The resistance also gives a *downward* acceleration of magnitude kv/m, so we can discard **A**. The key is therefore **C**.

Example 5 The velocity of P is $2\mathbf{i}\,\mathrm{m\,s}^{-1}$ and the velocity of Q is $(2\sqrt{3})\mathbf{j}\,\mathrm{m\,s}^{-1}$. A unit vector in the direction of the velocity of P relative to Q is

A $\mathbf{i} - \sqrt{3}\mathbf{j}$ **B** $2(\mathbf{i} - \sqrt{3}\mathbf{j})$ **C** $2(-\mathbf{i} + \sqrt{3}\mathbf{j})$

D $\frac{1}{2}(\mathbf{i} - \sqrt{3}\mathbf{j})$ **E** $\frac{1}{2}(-\mathbf{i} + \sqrt{3}\mathbf{j})$

A, B, C can be discarded at once because they do not give unit vectors (with modulus unity). The velocity of P relative to Q is $\mathbf{V}_P - \mathbf{V}_Q = 2(\mathbf{i} - \sqrt{3}\mathbf{j})\,\mathrm{m\,s}^{-1}$ and only **D** of the remaining **D** and **E** gives a vector in this direction. Hence the key is **D**.

(5) Identifying some important points or sometimes a partial working of the problem will enable one to identify the correct key.

Example 6 A small ring P, of mass m, is threaded on a smooth circular wire of centre O fixed in a vertical plane. The ring is projected from the lowest point and just reaches a point Q of the wire, level with O. When P is at Q the magnitude of the force exerted by the wire on the ring

A   cannot be found from the given data

B   is 0

C   is mg

D   is $2mg$

E   is $3mg$

When at Q, the weight of the bead mg acts vertically downwards (along the tangent to the wire). Also at Q the mv^2/r force is zero because $v = 0$ at Q. The key, therefore, is **B**.

Example 7 The force $\mathbf{F}$, where $\mathbf{F} = (\mathbf{i} + \mathbf{j} + 7\mathbf{k})\,\mathrm{N}$, acts at the point with position vector $(\mathbf{i} - 2\mathbf{j} + 8\mathbf{k})$ m. The moment of $\mathbf{F}$, in N m, about the point with position vector $(\mathbf{i} + 2\mathbf{k})$ m is

A 40 B 15 C $-2\mathbf{i} + 5\mathbf{j} + \mathbf{k}$

D $-20\mathbf{i} + 6\mathbf{j} + 2\mathbf{k}$ E $20\mathbf{i} - 6\mathbf{j} - 2\mathbf{k}$

As the moment of a force is a vector product we can discard **A** and **B**. The correct vector product is $(-2\mathbf{j} + 6\mathbf{k}) \times (\mathbf{i} + \mathbf{j} + 7\mathbf{k})$, but there is no need to work the whole product out—the first term is $-20\mathbf{i}$ which is present only in **D**. Hence key **D**.

Example 8 At time t the position vector of a particle P is $\mathbf{a} + \mathbf{v}t$, where $\mathbf{a}, \mathbf{v}$ are constant vectors. The least distance of P from the origin of vectors occurs when $t =$

A $-\mathbf{a}/\mathbf{v}$ B $-\mathbf{v}/\mathbf{a}$ C $\mathbf{a} - \mathbf{v}$

D $-\mathbf{v}^2/(\mathbf{a} \cdot \mathbf{v})$ E $-(\mathbf{a} \cdot \mathbf{v})/\mathbf{v}^2$

The keys **A** and **B** are meaningless as we cannot divide one vector by another. As t is a scalar, **C**, which is a vector, must be incorrect. Also the dimensions of **D** are $(LT^{-1})^2/(L \cdot LT^{-1}) = T^{-1}$ and not those of time. Therefore key is **E**.

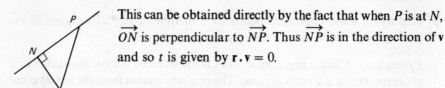

This can be obtained directly by the fact that when P is at N, $\overrightarrow{ON}$ is perpendicular to $\overrightarrow{NP}$. Thus $\overrightarrow{NP}$ is in the direction of $\mathbf{v}$ and so t is given by $\mathbf{r} \cdot \mathbf{v} = 0$.

Note that in some cases key **E**, often written as 'none of the above', may be correct. It must not be automatically discarded.

Example 9 Power is measured in the same units as

A energy B linear momentum

C force D work

E none of the above

Power is a rate of doing work with dimensions ML^2T^{-3}. Energy and work are equivalent (ML^2T^{-2}). Force is MLT^{-2} and linear momentum is MLT^{-1}. The key is, therefore, **E**.

Multiple-completion

In a multiple-completion question ONE or MORE of the responses given are correct. You have to decide which of the responses is (are) correct. Then choose

A   if **1**, **2** and **3** are correct

B   if only **1** and **2** are correct

C   if only **2** and **3** are correct

D   if only **1** is correct

E   if only **3** is correct

Directions summarized			
A	1	2	3
B	1	2	
C		2	3
D	1		
E			3

This appears to mean that three separate problems are involved in these questions. However, the three statements made will usually have a common theme, for instance, features of a simple harmonic motion or various probabilities deduced from common data, and often a little work done on the data will give us all the answers. This is best illustrated by examples.

Example 10   A bag contains just 10 green balls, 6 blue balls and 4 red balls. A ball is to be drawn out at random and then returned to the bag. A second ball is then to be drawn out at random. The probability that

1   the first ball will be red and the second will be blue is $\frac{3}{50}$

2   both balls will be green is $\frac{1}{4}$

3   neither ball will be red is $\frac{16}{25}$

The results of the first and second drawings are independent and are

$$P(G) = \tfrac{10}{20} = \tfrac{1}{2}, \qquad P(B) = \tfrac{6}{20} = \tfrac{3}{10}, \qquad P(R) = \tfrac{4}{20} = \tfrac{1}{5}.$$

Also $P(R') = 1 - P(R) = \tfrac{4}{5}$. The required probabilities can then be worked mentally and ticked off as you do them:

$$P(R, B) = \tfrac{1}{5} \times \tfrac{3}{10} \qquad \text{so } \mathbf{1} \text{ is correct,}$$

$$P(G, G) = \tfrac{1}{2} \times \tfrac{1}{2} \qquad \text{so } \mathbf{2} \text{ is correct,}$$

$$P(R', R') = \tfrac{4}{5} \times \tfrac{4}{5} \qquad \text{so } \mathbf{3} \text{ is correct.}$$

The key is, therefore, **A**.

Example 11 At time $t = 0$ a particle P is projected from O with horizontal and vertical components of velocity of 20 m s^{-1} and 10 m s^{-1} respectively. $g = 10 \text{ m s}^{-2}$. When $t = 1$ s, P

1 is moving horizontally

2 is more than 20 m from O

3 has a speed less than 20 m s^{-1}

Here a figure is helpful.
Vertical motion—when $t = 1$, $v = 10 - 10 = 0$.
Horizontal motion—the speed is constant $= 20 \text{ m s}^{-1}$.
All the answers can be decided from these two facts.
When $t = 1$, vertical speed is zero so **1** is correct.
When $t = 1$, P has travelled 20 m horizontally and some distance vertically and so is more than 20 m from O (no need to work out the vertical distance), so **2** is correct.
When $t = 1$, the speed is exactly 20 m s^{-1} so **3** is incorrect.
Hence key is **B**.

Example 12 The displacement x m of a particle moving on the x-axis is given, at time t seconds, by

$$x = 3 \sin (2\pi t/7).$$

1 When $t = 0$, its speed is 3 m s^{-1}

2 When $x = 3$, its speed is zero

3 The motion is periodic with period 7 s

The work we need to do here is to find the velocity, $v \text{ m s}^{-1}$, by differentiation, i.e.

$$v = \frac{\mathrm{d}x}{\mathrm{d}t} = \frac{6\pi}{7} \cos (2\pi t/7).$$

Then, when $t = 0$, $v = 6\pi/7$ so **1** is incorrect.
When $x = 3$, $\sin (2\pi t/7) = 1 \Rightarrow \cos (2\pi t/7) = 0 \Rightarrow v = 0$, so **2** is correct.
Finally $3 \sin (2\pi t/7)$ is a periodic function with period $2\pi/(2\pi/7) = 7$ so **3** is correct.
Hence key is **C**.

However (see the note following this example), **2** alone is not a key so there was no real need to consider **3** as it must be true.

Note that it is worth observing that not all the logical possibilities are present in the keys. Thus, **2** only or **1** and **3** only are not possible keys. This means that when **1** is incorrect then **3** must be correct.

If **1** is correct, key cannot be **C** or **E**.
If **1** is incorrect and **2** is incorrect, key must be **E**.
If **1** is incorrect and **2** is correct, key must be **C**.

Example 13   A particle P, of mass 1 kg, moves along Ox and performs simple harmonic motion of amplitude 6 m under the action of a force $4|x|$ N directed towards O.

1   The period of the motion is $\pi/2$ s

2   The maximum speed is of magnitude 12 m s^{-1}

3   The maximum acceleration is of magnitude 24 m s^{-2}

The general equation of motion for simple harmonic motion (SHM) is $\ddot{x} = -\omega^2 x$ and all that is necessary for this question is to find the value of ω.
   As the equation of motion in this case is $\ddot{x} = -4x$, it follows that $\omega = 2$.
Then, period $= 2\pi/\omega = \pi$ and so **1** is incorrect.
Maximum speed $= \omega a = 12$ m s^{-1} and so **2** is correct.
It follows, by the previous note, that **3** is correct and the key is **C**. However, as a check, maximum accleration $= \omega^2 a = 24$ m s^{-2}, which is given in **3**.

Example 14

The uniform beam PQR, of mass 20 kg and length 6 m, rests horizontally in equilibrium, on supports at Q and R, where $PQ = 1$ m. ($g = 10$ m s^{-2}.)

1   The force exerted by the beam on the support at Q is of magnitude 180 N

2   The force exerted by the beam on the support at R is of magnitude 80 N

3   The magnitude of the greatest load which can be hung from P without disturbing equilibrium is 400 N

Clearly, taking moments about R and Q in turn and remembering that the weight of the beam is of magnitude $20g$ N ($= 200$ N) we find that **1** is incorrect and **2** is correct. It follows that **3** must be correct and the key is **C**.
Checking for **3**, taking moments about Q: $W \times 1 = 200 \times 2$ N $\Rightarrow W = 400$ N.

Example 15   In t seconds from starting a wheel rotates through θ radians, where

$$\theta = 12t - 4t^2.$$

1   When $t = 0$, the angular speed is 12 rad s^{-1}.

2 When $t = 1\frac{1}{2}$, the angular acceleration is zero

3 The wheel rotates through 24 radians in 2 s

We need angular speed and angular acceleration here and these can be found by successive differentiations.

$$\frac{d\theta}{dt} = 12 - 8t,$$

$$\frac{d^2\theta}{dt^2} = -8.$$

When $t = 0$, $\dfrac{d\theta}{dt} = 12$ rad s^{-1} and so **1** is correct.

When $t = 1\frac{1}{2}$, $\dfrac{d^2\theta}{dt^2} = -8$ rad s^{-2} and so **2** is incorrect

It follows that **3** is incorrect and the key is **D**.
As a check, when $t = 2$, $\theta = 24 - 16 = 8$ and so **3** is incorrect.

5.2   Paper 3   (multiple-choice)

Time allowed: $1\frac{1}{4}$ hours

The numerical value of g should be understood to be available for questions in any section if required. Take g as 10 m s^{-2} unless otherwise given.

SECTION I

Questions 1–20   **(Twenty questions)**

Directions. Each of these questions is followed by five suggested answers. For each question, select the answer which you think is correct.

1   The velocity v m s^{-1} of a point P moving along Ox is given by $v = e^{-x/2}$, where $OP = x$ m. Its acceleration f m s^{-2} is given by

A $v = -2f$ **B** $v^2 = -2f$ **C** $v^2 = 2f$

D $v^2 = 2fx$ **E** $v^2 = -2fx$

2   A train, of mass m, moves with acceleration a along a straight level track against a constant resistance of magnitude R. When the speed is v, the rate at which the engine of the train is working is

A Rv **B** $(R - ma)v$ **C** $(ma - R)v$

D $(R + ma)v$ **E** $(R + ma)v^2/2$

3   A projectile is projected with velocity V at an angle α to the horizontal. Its greatest height above the horizontal plane through the point of projection is

A $(V^2 \sin 2\alpha)/g$ **B** $(V^2 \sin 2\alpha)/(2g)$

C $(V^2 \sin^2 \alpha)/(2g)$ **D** $(V^2 \sin^2 \alpha)/g$

E $(V^2 \cos^2 \alpha)/(2g)$

4

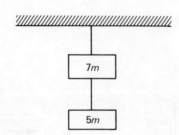

The system hangs in equilibrium. The tensions in the upper and lower strings are respectively

A $7mg, \ 5mg$ **B** $12mg, \ 12mg$ **C** $7mg, \ 2mg$

D $12mg, \ 6mg$ **E** $12mg, \ 5mg$

5   A wheel makes 1 revolution every four seconds. The angular speed of the wheel, in rad s^{-1}, is

A 2π **B** π **C** $\pi/2$ **D** $1/\pi$ **E** $1/(2\pi)$

6   A green die and a white die are to be rolled together. The probability that the white die will show a larger number than the green die is

A $\frac{1}{2}$ **B** $\frac{1}{3}$ **C** $\frac{2}{5}$ **D** $\frac{5}{12}$

E none of the above

7

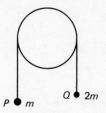

P ● m   Q ● 2m

Two particles P and Q, of masses m and $2m$ respectively, are connected by a light inextensible string passing over a fixed, smooth light pulley. The system is released from rest with the string taut and the hanging parts vertical. In the subsequent motion the force exerted by the string on the pulley is of magnitude

A $8mg/3$ **B** $4mg/3$ **C** mg **D** $3mg$

E none of the above

8

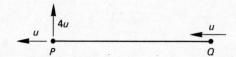

Two particles P and Q, of masses m and $2m$, respectively, are connected by a light inextensible rod and rest on a smooth horizontal table. An impulse given to P gives the particles the velocities shown. The magnitude of the impulse is

A $5mu$ **B** $3mu$ **C** $2mu$

D $\sqrt{17}mu$ **E** $(1 + \sqrt{17})mu$

9 A small bead, of mass m, is threaded on a smooth wire which is bent into a circle, of radius r, and which is fixed in a vertical plane. The bead is projected from the lowest point of the wire with speed $\sqrt{(5rg)}$. The force exerted by the bead on the wire at the highest point is

A mg upwards   **B** mg downwards   **C** $3mg$ upwards

D $3mg$ downwards   **E** zero

10

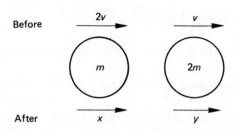

Before
2v
v
m
2m
After
x
y

Two perfectly elastic spheres collide directly as shown.

A $x = v, y = 2v$ **B** $x = -2v, y = 3v$

C $x = -5v/3, y = 2v/3$ **D** $x = 5v/3, y = 2v/3$

E $x = 2v/3, y = 5v/3$

11 The velocity of P relative to Q is $(-2\mathbf{i} + 3\mathbf{j} - \mathbf{k})$ m s^{-1} and the velocity of P is $(\mathbf{i} - 4\mathbf{j} + \mathbf{k})$ m s^{-1}. The velocity of Q is

A $(3\mathbf{i} - 7\mathbf{j} + 2\mathbf{k})$ m s^{-1} **B** $(-\mathbf{i} - \mathbf{j})$ m s^{-1}

C $(\mathbf{i} + \mathbf{j})$ m s^{-1} **D** $(-3\mathbf{i} + 7\mathbf{j} - 2\mathbf{k})$ m s^{-1}

E none of the above

12 A particle of mass m moves under the action of a constant force $\mathbf{F}$. At time $t = 0$ the particle is at the origin of position vectors and is moving with velocity $\mathbf{V}$. Its position vector at time t is

A $(m\mathbf{V} + \mathbf{F})t/m$ **B** $(m\mathbf{V}t + \mathbf{F})t/(2m)$

C $(2m\mathbf{V} + \mathbf{F}t)t/(2m)$ **D** $(m\mathbf{V} + \tfrac{1}{2}\mathbf{F})t^2/m$

E none of the above

13

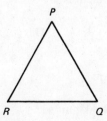

PQR is an isosceles triangle.
The resultant of forces represented by $\overrightarrow{PQ}$, $\overrightarrow{QR}$ and $\overrightarrow{RP}$ is a

A single force represented by $2\overrightarrow{PQ}$

B single force through R parallel to PQ and of magnitude $2|\overrightarrow{PQ}|$

C single force represented by $2\overrightarrow{PX}$, where X is the mid-point of RQ

D couple whose moment is of magnitude $\triangle PQR$

E couple whose moment is of magnitude $2\triangle PQR$

14

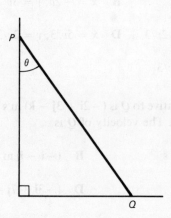

The uniform plank PQ rests in limiting equilibrium with its upper end P against a smooth vertical wall and its lower end Q on a rough horizontal plane. The coefficient of friction between the plank and the wall is

A $\frac{1}{2}\cot\theta$ **B** $\frac{1}{2}\tan\theta$ **C** $2\tan\theta$

D $2\cot\theta$ **E** $\frac{1}{2}\sin\theta$

15 The independent probabilities that Anne, Brenda and Carol will solve a particular problem are $\frac{1}{2}, \frac{1}{3}$ and $\frac{1}{4}$ respectively. The probability that just two of them will solve this problem is

A $\frac{3}{8}$ **B** $\frac{1}{8}$ **C** $\frac{1}{12}$ **D** $\frac{1}{4}$ **E** $\frac{1}{6}$

16 The dimensions of work are

A kg m s^{-2} **B** $\text{kg m}^2\text{s}^{-1}$ **C** $\text{kg m}^2\text{s}^{-2}$

D kg m s^{-1} **E** $\text{kg m}^2\text{s}^{-3}$

17 The natural length of an elastic spring which obeys Hooke's Law is 3 m. When the tension in the string is 50 N, the length of the spring is 3·6 m. The potential energy, in J, stored in the spring is

A 150 **B** 90 **C** 45 **D** 30 **E** 15

18

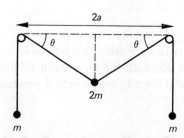

Three particles are attached as shown to a light inextensible string, one at each end and one at its mid-point. When the string hangs over two small smooth pegs at the same level and at a distance $2a$ apart, with the inclined parts each making an angle θ with the horizontal and the other parts of the string vertical, the potential energy of the system can be expressed as

A $2mga(\sec \theta - \tan \theta) + \text{constant}$

B $2mga(\sec \theta + \tan \theta) + \text{constant}$

C $-2mga(\sec \theta - \tan \theta) + \text{constant}$

D $-2mga(\sec \theta + \tan \theta) + \text{constant}$

E none of the above

19 A particle of mass m is on the floor of a lift which, at a particular instant, is moving upwards with speed v and downward acceleration f, where $f < g$. The upward force exerted by the floor of the lift on the particle is of magnitude

A mfv **B** $m(f + g)$ **C** $m(f + g)v$

D $m(g - f)$ **E** $m(g - f)v$

20

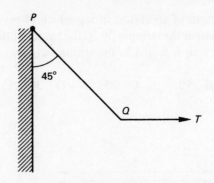

A uniform rod PQ, of mass 12 kg, is smoothly hinged to a vertical wall at P. The rod makes the angle 45° with the downward vertical and is kept in equilibrium by a horizontal force of magnitude T newtons acting at Q.

$T =$

A $3g$ **B** $6g$ **C** $12g$ **D** $6\sqrt{2}g$ **E** $3\sqrt{2}g$

SECTION II

Questions 21–30   **(Ten questions)**

Directions. For each of the following questions, **ONE** or **MORE** of the responses given are correct. Decide which of the responses is (are) correct. Then choose

		Directions summarized		
A	if **1, 2** and **3** are correct	**A**	**1**	**2**	**3**
B	if only **1** and **2** are correct	**B**	**1**	**2**	
C	if only **2** and **3** are correct	**C**		**2**	**3**
D	if only **1** is correct	**D**	**1**		
E	if only **3** is correct	**E**			**3**

21 X and Y are independent events. $P(X) = \frac{1}{4}$, $P(Y) = \frac{1}{3}$.

1 $P(X \cap Y) = \frac{1}{12}$ **2** $P(X|Y) = \frac{1}{4}$ **3** $P(X' \cap Y') = \frac{11}{12}$

22 The position vector **r**, referred to a fixed origin O, of a particle at time t is given by

$$\mathbf{r} = a(\mathbf{i} \sin \omega t - \mathbf{j} \cos \omega t),$$

where a, ω are non-zero constants.

1 $|\mathbf{r}| = a$

2 $\dfrac{d^2 \mathbf{r}}{dt^2} = -\omega^2 \mathbf{r}$

3 $\dfrac{d\mathbf{r}}{dt}$ is perpendicular to **r**

23 A particle slides down the outside of a fixed smooth sphere.

1 The energy of the particle remains constant

2 The acceleration of the particle is of magnitude g

3 The particle leaves the sphere when it is level with the centre of the sphere.

24 An elastic sphere of mass m falls vertically and strikes a horizontal plane with speed v. The sphere rebounds with speed u.

1 The coefficient of restitution between the sphere and the plane is v/u.

2 The upward impulse given to the sphere on impact is of magnitude $m(v - u)$

3 The height above the plane to which the sphere rebounds is $u^2/(2g)$

25 A constant force, of magnitude 18 N, acts on a body of mass 3 kg. At a particular instant the speed of the body is 6 m s^{-1}. At this instant

1 the kinetic energy of the body is 54 J

2 the power developed by the force is 108 W

3 the acceleration of the body is 6 m s^{-2}

26 A particle P, of mass 1 kg, performs simple harmonic motion of amplitude 2 m along Ox, under the action of a force of magnitude $9|x|$ N directed towards O. Then

1 the period of the motion is $2\pi/9$ s

2 the maximum speed of P is 18 m s^{-1}

3 the greatest magnitude of the acceleration of P is 18 m s^{-2}

27

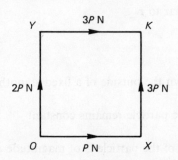

$OXKY$ is a square. The resultant of the forces shown

1 passes through the mid-point of KX **2** passes through O

3 is of magnitude $(P\sqrt{41})$ N

28 At time $t = 0$, two particles P_1 and P_2 start from O and move along Ox so that their displacements, x_1 m and x_2 m respectively, at time t seconds are given by $x_1 = t^2$, $x_2 = t$.

1 At first P_1 moves ahead of P_2

2 P_1 and P_2 have equal speeds when $t = \frac{1}{2}$

3 P_1 and P_2 coincide for just one positive value of t

29 Three coplanar non-zero, non-parallel, non-collinear forces $\mathbf{P}, \mathbf{Q}$ and $\mathbf{R}$ are in equilibrium.

1 The lines of action of the forces are concurrent

2 $|\mathbf{P}| < |\mathbf{Q}| + |\mathbf{R}|$

3 $\mathbf{P} \cdot \mathbf{Q} = \mathbf{Q} \cdot \mathbf{R} = \mathbf{R} \cdot \mathbf{P}$

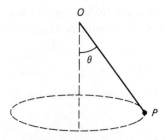

The diagram shows a conical pendulum in which a particle P, of mass m, moves in a horizontal circle with speed v. The string is of length l, it makes an angle θ with the downward vertical and the tension in the string is T.

1 $T = mg \sec \theta$ **2** $T \sin \theta = mv^2/l$

3 The period of revolution is $2\pi l/v$

5.3 Solutions for Paper 3

1 Acceleration can be obtained by differentiation, but notice that v is given in terms of x *not* t. Hence

$$f = v\frac{dv}{dx} = v \cdot (-\tfrac{1}{2}e^{-x/2}) = -\tfrac{1}{2}v^2, \qquad \text{giving key } \mathbf{B}.$$

2 Let the pull of the engine $= P$ N. Then $P = R + ma$.
Rate of work = work done per second $= (R + ma)v$, giving key **D**.

3 Many students know this result; otherwise, consider the vertical motion and use

$$\text{`}v^2 = u^2 + 2as\text{'} \quad \Rightarrow \quad 0 = V^2 \sin^2 \alpha - 2gs, \qquad \text{giving key } \mathbf{C}.$$

4 Upper string supports *both* bodies, so tension $= 12mg$.
Lower string supports only lower body, so tension $= 5mg$, giving key **E**.

5 1 revolution $= 2\pi$ radians $\quad \Rightarrow \quad$ angular speed $= \pi/2$ rad s^{-1}, giving key **C**.

6   When 2 dice are rolled, there are 36 different outcomes of which 6 give equal scores. Of the remaining 30, there must be 15 in which the white score > the green score, giving

$$P(W > G) = \tfrac{15}{36} = \tfrac{5}{12}, \qquad \text{giving key } \mathbf{D}.$$

OR   P(equal scores) $= \tfrac{1}{6}$;   P(unequal scores) $= \tfrac{5}{6}$; $P(W > G) = \tfrac{5}{12}$.

7   The principles involved here are:

(a) a fixed pulley means that the pressure on the pulley is equal to the sum of the tensions in the two parts of the string,

(b) smooth pulley means that these two tensions are equal,

(c) considering the motion of each particle separately

$$\left.\begin{array}{r} 2mg - T = 2ma \\ T - mg = ma \end{array}\right\} \;\Rightarrow\; T = 4mg/3, \qquad \text{giving key } \mathbf{A}.$$

8   Use the momentum–impulse law, i.e.

$$\text{Impulse} = \text{momentum communicated.}$$

Along QP the impulse is $3mu$.
Perpendicular to QP the impulse is $4mu$.

The magnitude of the impulse $= \sqrt{(9 + 16)}mu = 5mu$, giving key $\mathbf{A}$.

9   (a) Use the principle of conservation of energy to get the speed V at the top, i.e.

$$\tfrac{1}{2}m \cdot 5gr - \tfrac{1}{2}mV^2 = mg \cdot 2r \;\Rightarrow\; V^2 = gr.$$

(b) Net downward force on bead at top $= mg + R = mV^2/r = mg \Rightarrow R = 0$, giving key $\mathbf{E}$.

10   By the conservation of momentum,

$$4mv = mx + 2my.$$

By Newton's Experimental Law,

$$y - x = v$$

$$\Rightarrow\; y = 5v/3, \quad x = 2v/3, \qquad \text{giving key } \mathbf{E}.$$

Note that the value of y only will give you the key, although it is as well to check x.

11 $V_P - V_Q = (-2i + 3j - k)\,m\,s^{-1}$
$\qquad V_P = (i - 4j + k)\,m\,s^{-1}$ $\quad\Big\}\quad \Rightarrow\quad V_Q = (3i - 7j + 2k)\,m\,s^{-1},$

giving key **A**.

12 $F = ma$, giving $a = F/m$.
Constant force means constant acceleration in a straight line, and so the equations of constant acceleration may be used.

$$\text{Using} \quad 's = ut + \tfrac{1}{2}at^2',$$

$$s = Vt + \tfrac{1}{2}(F/m)t^2, \qquad \text{giving key } \mathbf{C}.$$

This may also be done by integration of $\dfrac{dv}{dt} = F/m$.

13 The key is **E** and this is a theorem of statics, but it could be obtained thus:

Moments of forces about $R = PQ \times$ perpendicular distance from R to PQ

$$= 2\triangle PQR, \qquad \text{clockwise}.$$

Similarly, moments about P and Q are also equal to $2\triangle PQR$, clockwise. Hence the resultant of the system is a couple of this moment.

14 By resolving and taking moments we get

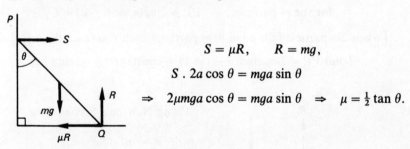

$$S = \mu R, \qquad R = mg,$$

$$S \cdot 2a \cos \theta = mga \sin \theta$$

$$\Rightarrow\quad 2\mu mga \cos \theta = mga \sin \theta \quad \Rightarrow \quad \mu = \tfrac{1}{2} \tan \theta.$$

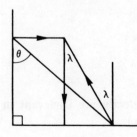

Alternatively, using angle of friction and the lines of action of three forces meeting in a point,

$$\tan \lambda = \tfrac{1}{2} \tan \theta$$

$$\Rightarrow \quad \mu = \tfrac{1}{2} \tan \theta, \qquad \text{giving key } \mathbf{B}.$$

15 The probabilities are independent so the product rule operates. When just two solve the problem, the other one does not.

$$\text{P}(A \text{ and } B \text{ solve, } C \text{ fails}) = \tfrac{1}{2} \times \tfrac{1}{3} \times \tfrac{3}{4}.$$

Likewise for the other combinations, and so

$$\text{P}(2 \text{ solve, } 1 \text{ fails}) = \tfrac{1}{2} \times \tfrac{1}{3} \times \tfrac{3}{4} + \tfrac{1}{3} \times \tfrac{1}{4} \times \tfrac{1}{2} + \tfrac{1}{2} \times \tfrac{1}{4} \times \tfrac{2}{3}$$

$$= \tfrac{6}{24} = \tfrac{1}{4}, \qquad \text{giving key } \mathbf{D}.$$

16 Work = force × distance
$$= \text{mass} \times \text{acceleration} \times \text{distance}$$
$$= \text{kg} \times \text{m s}^{-2} \times \text{m}, \qquad \text{giving key } \mathbf{C}.$$

17 Potential energy in spring = mean tension × extension

$$= \left(\frac{0 + 50}{2}\right) \times 0.6 \text{ J}$$

$$= 15 \text{ J}, \qquad \text{giving key } \mathbf{E}.$$

18 Refer to a fixed horizontal level, in this case the level of the pulleys;

for the $2m$ particle, $\qquad \text{PE} = -2mga \tan \theta + C_1,$

for the m particles, $\qquad \text{PE} = 2mg(a \sec \theta - a) + C_2$

$\left[\text{when } 2m \text{ particle falls } a \tan \theta, \ m \text{ particles each rise } (a \sec \theta - a)\right].$

Total PE $= 2mga(\sec \theta - \tan \theta) + \text{constant}, \qquad$ giving key $\mathbf{A}$.

19

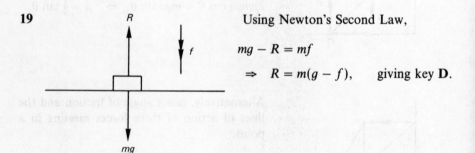

Using Newton's Second Law,

$$mg - R = mf$$

$$\Rightarrow \quad R = m(g - f), \qquad \text{giving key } \mathbf{D}.$$

(Note that the direction of motion and velocity are irrelevant in this question; force depends only on acceleration.)

20 This is best done by taking moments about P, thus excluding the force at the hinge.

Let $PQ = 2a$, then

$$T \cdot 2a \cos 45° = 12ga \sin 45°$$

$$\Rightarrow \quad T = 6g, \quad \text{giving key } \mathbf{B}.$$

21 Independent events $\Rightarrow P(X \cap Y) = P(X) \cdot P(Y)$, hence **1** is correct.
Independent events $\Rightarrow P(X \mid Y) = P(X)$, hence **2** is correct,

$$P(X' \cap Y') = P(X') \cdot P(Y') = \tfrac{3}{4} \cdot \tfrac{2}{3},$$

hence **3** is incorrect, giving key **B**.

22 **1** $|\mathbf{r}| = a\sqrt{(\sin^2 \omega t + \cos^2 \omega t)} = a$, and so **1** is correct.

2 $\dfrac{d(\mathbf{r})}{dt} = a\omega(\mathbf{i} \cos \omega t + \mathbf{j} \sin \omega t)$,

$\dfrac{d^2(\mathbf{r})}{dt^2} = a\omega^2(-\mathbf{i} \sin \omega t + \mathbf{j} \cos \omega t) = -\omega^2 \mathbf{r}$ and so **2** is correct.

3 $\dfrac{d\mathbf{r}}{dt} \cdot \mathbf{r} = a^2\omega(\sin \omega t \cos \omega t - \cos \omega t \sin \omega t) = 0$ and so **3** is correct.

Alternatively, this might be recognized as motion in a circle of radius a with constant angular speed ω. Then all the options follow immediately. Hence key **A**.

23 **1** is correct as conservation of energy applies.

2 Particle is not falling freely but is subject to a reaction from the sphere so **2** is incorrect.

3 Particle leaves the sphere when R disappears which must happen before the radial component of the weight disappears so **3** is incorrect, giving key **D**.

Note that since **1** is correct and **2** is incorrect, then **3** must be incorrect.

24 In the simple case of a vertical rebound

$$u = ev,$$

where e is the coefficient of restitution,

$$\Rightarrow \quad e = u/v,$$

hence **1** is incorrect.

For **2**, use the fact that the upward impulse is equal to the change in the upward momentum $= mu - (-mv) = m(u + v)$, and so **2** is incorrect.

It follows, then, that **3** must be correct, but checking by conservation of energy,

$$\tfrac{1}{2}mu^2 = mgh,$$

where h is the height reached,

$$\Rightarrow \quad h = u^2/(2g), \qquad \text{giving key } \mathbf{E}.$$

25   **1** $\text{KE} = \tfrac{1}{2} \times 3 \times 36\,\text{J} = 54\,\text{J}$ so **1** is correct.

     2   Power = work done per second $= 18 \times 6\,\text{W} = 108\,\text{W}$ so **2** is correct.

     3 $18 = 3a$ (Newton's Second Law) $\Leftrightarrow a = 6$, so **3** is correct, giving key **A**.

26   General equation for SHM is $\ddot{x} = -\omega^2 x \Rightarrow \omega = 3\,\text{s}^{-1}$.

     1   Period $= 2\pi/\omega = 2\pi/3$ s, so **1** is incorrect.

     2   Maximum speed $= \omega a = 6\,\text{m s}^{-1}$, so **2** is incorrect. It follows that **3** is correct but, checking, greatest acceleration $= 9 \times 2\,\text{m s}^{-2} = 18\,\text{m s}^{-2}$, so **3** is correct, giving key **E**.

27   **1**   Taking moments about the mid-point of KX, we get

$$4Pa + 3Pa - Pa \neq 0,$$

so the resultant does not pass through this point and **1** is incorrect.

     2   Taking moments about O,

$$6Pa - 6Pa = 0,$$

so the resultant does pass through this point and **2** is correct. Hence **3** is correct and the key is **C**.

However, checking for **3**, resultant is of magnitude

$$\sqrt{(25 + 16)}P = \sqrt{41}P.$$

28 The speeds of P_1 and P_2 are given by $\dfrac{dx_1}{dt} = 2t$ and $\dfrac{dx_2}{dt} = 1$.

1 Initially the speeds are 0 and 1 m s^{-1}, respectively, so **1** is incorrect.

2 When $t = \frac{1}{2}$, both speeds $= 1\text{ m s}^{-1}$ so **2** is correct.

It follows that **3** is correct and the key is **C**.
However, checking for **3**, $x_1 = x_2$ when $t = 1$ so the particles coincide at this time.

29 1 is correct as the forces are non-zero and non-parallel.

2 For equilibrium, the forces are proportional in magnitude to the sides of a triangle and, as any side of a triangle $<$ the sum of the other two sides, $|\mathbf{P}| < |\mathbf{Q}| + |\mathbf{R}|$, and so **2** is correct.

3 is incorrect—there is no reason why this relation should exist and it is easily disproved by considering the case when $\mathbf{P}$ and $\mathbf{Q}$ are perpendicular, when $\mathbf{P} \cdot \mathbf{Q} = 0$, giving key **B**.

30 1 Resolving vertically $T \cos \theta = mg$ so **1** is correct.

2 The acceleration towards the centre is $v^2/(l \sin \theta)$ so $T \sin \theta = mv^2/(l \sin \theta)$ and **2** is incorrect.

It follows that **3** is incorrect and the key is **D**.
Checking for **3**, circumference of the circle $= 2\pi l \sin \theta$ and the period $= (2\pi l \sin \theta)/v$ and so **3** is incorrect.

5.4   Paper 4 (multiple-choice)

Time allowed: $1\frac{1}{4}$ hours

SECTION I

Questions 1–20   **(Twenty questions)**

Directions. Each of these questions is followed by five suggested answers. For each question, select the answer which you think is correct.

1 A projectile is given an initial velocity V at an angle α to the horizontal. Its range on the horizontal plane through the point of projection is

A $(V^2 \sin 2\alpha)/(2g)$ **B** $(V^2 \sin 2\alpha)/g$

C $(V^2 \cos 2\alpha)/g$ **D** $(V^2 \sin^2 \alpha)/(2g)$

E $(V^2 \sin^2 \alpha)/g$

2 The probability that Eva will solve a problem is $\frac{1}{2}$ while for Mary in an independent attempt the probability is $\frac{1}{3}$. The probability that both Eva and Mary will solve the problem when working independently is

A $\frac{5}{6}$ **B** $\frac{2}{3}$ **C** $\frac{5}{12}$ **D** $\frac{1}{6}$ **E** $\frac{1}{12}$

3 Impulse is measured in the same units as

A force   **B** work   **C** energy

D linear momentum   **E** power

4

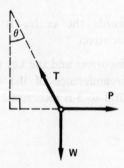

A load, of weight **W**, is attached to the end of a light inextensible string, inclined at the angle θ to the downward vertical and is maintained in equilibrium by a horizontal force **P**. The tension in the string is **T**. Then $\tan \theta =$

A $|T|/|W|$ **B** $|P|/|T|$ **C** $|P|/|W|$

D $|W|/|P|$ **E** $|W|/|T|$

5 A particle moves along Ox so that its displacement x m at time t seconds is given by

$$x = \sin t + \cos t + \frac{t\sqrt{3}}{2}.$$

The velocity, in m s^{-1}, when $t = \pi/3$, is

A $\frac{1}{2}$ **B** $-\frac{1}{2}$ **C** $\frac{1}{2} + \sqrt{3}$

D $-\frac{1}{2} + \sqrt{3}$ **E** $-\frac{1}{2} + \sqrt{3}/2 + \sqrt{3\pi^2/36}$

6 The tension in an elastic string, of modulus λ and natural length l, which is stretched to a length $(l + a)$ is

A $\lambda(l + a)/l$ **B** $\lambda a/(l + a)$ **C** $\lambda a/l$

D $\lambda l/a$ **E** $\lambda(l + a)/a$

7 A particle, of mass 0·1 kg and initially at rest, is acted upon by a constant force of magnitude 2 N for one minute. Its final speed, in m s^{-1}, is

A 0·2 **B** 12 **C** 20 **D** 120 **E** 1200

8

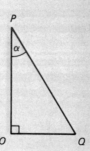

A uniform plank PQ rests in limiting equilibrium at an acute angle α to the downward vertical with the end P against a smooth vertical wall and the end Q on rough horizontal ground. The coefficient of friction between the plank and the ground is

A $2 \cot \alpha$ **B** $\frac{1}{2} \tan \alpha$ **C** $\frac{1}{2}$

D $2 \tan \alpha$ **E** $\cot \alpha$

9 $P(X \cap Y) = \frac{1}{4}$, $P(Y) = \frac{1}{3}$.

$P(X|Y) =$

A $\frac{1}{3}$ **B** $\frac{1}{4}$ **C** $\frac{1}{12}$ **D** $\frac{4}{3}$ **E** $\frac{3}{4}$

10 Two elastic strings XY and YZ, each of natural length l, are joined at Y. The end X is fixed and a weight $\mathbf{W}$ is attached at Z. The strings XY and YZ would be doubled in length by tensions $2\mathbf{W}$ and $4\mathbf{W}$ respectively. In equilibrium, $XZ =$

A $2\frac{3}{4}l$ **B** $2\frac{3}{8}l$ **C** $2\frac{1}{6}l$ **D** $2\frac{1}{3}l$ **E** $2\frac{2}{3}l$

11 A point in simple harmonic motion takes $\pi/2$ seconds to travel from one end of its path to the other, a distance of 4 m. Its maximum acceleration, in $\mathrm{m\,s^{-2}}$, is

A 1   **B** 2   **C** 4   **D** 8   **E** 16

12 The magnitude of the acceleration of a lorry of mass m and power P, when travelling at speed v along a straight level road against a resistance R, is

A $(P - vR)/(mv)$ **B** $(P - vR)/m$ **C** $(Pv - R)/m$

D $(Pv - R)/(mv)$ **E** P/m

13

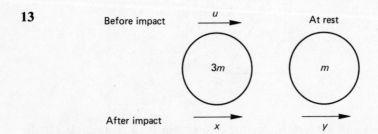

Given that the coefficient of restitution between the spheres is $\frac{1}{3}$, then $x =$

A $u/3$ **B** $u/2$ **C** $2u/3$ **D** $5u/6$ **E** $4u/3$

14 The force exerted on a body of mass m, when at a distance R from the centre of the Earth, is of magnitude GMm/R^2, where M is the mass of the Earth. In the SI system of units the dimensions of G are

A $\ \text{kg}^{-1}\,\text{m}^3\,\text{s}^{-1}$

B $\ \text{kg}^{-1}\,\text{m}^3\,\text{s}^{-2}$

C $\ \text{kg}^{-1}\,\text{m}^2$

D $\ \text{kg}^{-2}\,\text{m}^2$

E $\ $ none of the above

15

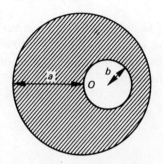

The diagram shows two circles of radii a and b, where $b < a/2$ and the centre O of the larger circle lies on the circumference of the smaller circle. The distance of the centroid of the shaded region from O is

A $\ b^3/a^2$

B $\ b^2/(a - b)$

C $\ b^2/(a^2 - b^2)$

D $\ b^3/(a^2 - b^2)$

E $\ \pi b^3/(a^2 - b^2)$

16 A particle moves in a circle of radius 1 m with centre O. The angular speed of the radius OP at time t seconds is $4/(1 + t)$ rad s^{-1}. The acceleration of P when $t = 1$ is

A $\ 1\,\text{m s}^{-2}$ along OP produced

B $\ 1\,\text{m s}^{-2}$ in the direction of the velocity of P

C $\ 1\,\text{m s}^{-2}$ towards O

D $\ 1\,\text{m s}^{-2}$ in the opposite direction to the velocity of P

E $\ $ none of the above

17 A small smooth sphere of mass m strikes a wall. Before impact the velocity of the sphere is of magnitude u and is making an angle α with the normal to wall. The coefficient of restitution between the wall and the sphere is $\frac{1}{3}$. The impulse given to the sphere by the wall is of magnitude

A $(4mu \cos \alpha)/3$ **B** $(mu \cos \alpha)/3$ **C** $mu/3$

D $(mu \sin \alpha)/3$ **E** $(4mu \sin \alpha)/3$

18 Two particles P and Q, of masses $3m$ and $5m$ respectively, are attached to the ends of a light inextensible string which hangs over a smooth pulley. The system is released from rest with the hanging parts of the string taut and vertical. The tension in the string is

A $2mg$ **B** $15mg/4$

C $15mg/2$ **D** $8mg$

E none of the above

19 A point moves along Ox so that its displacement, x m, at time t seconds is given by $x^2 = t^2 + 1$. Its acceleration, in m s^{-2}, is

A $1/x$ **B** $1/x^3$ **C** $-t/x^2$

D $-t^2/x^3$ **E** $1/x - t/x^2$

20 A smooth ring P is threaded on a wire bent in the form of a circle of radius a and centre O. The circle rotates with constant angular speed ω about a vertical diameter AB, the ring remaining at rest relative to the wire at a distance $\frac{1}{2}a$ from AB.

$\omega^2 =$

A $2g/a$ **B** $g/(2a)$ **C** $2g/(a\sqrt{3})$

D $(g\sqrt{3})/(2a)$ **E** $2a/(g\sqrt{3})$

SECTION II

Questions 21–30 (Ten questions)

Directions. For each of the following questions, **ONE** or **MORE** of the responses given are correct. Decide which of the responses is (are) correct. Then choose

	Directions summarized		
A if **1**, **2** and **3** are correct	**A**	**1**	**2**	**3**
B if only **1** and **2** are correct	**B**	**1**	**2**	
C if only **2** and **3** are correct	**C**		**2**	**3**
D if only **1** is correct	**D**	**1**		
E if only **3** is correct	**E**			**3**

21 With the usual notation, the acceleration of a particle moving along Ox is given by

1 $\dfrac{dv}{dt}$ 　　　　 2 $\dfrac{d^2x}{dt^2}$ 　　　　 3 $v\dfrac{dv}{dx}$

22 Given that the two events X and Y are independent and $P(X) = 0\cdot2$, $P(Y) = 0\cdot5$, then

1 the probability that both X and Y will occur is $0\cdot1$

2 the probability that either X or Y or both will occur is $0\cdot7$

3 the probability that neither X nor Y will occur is $0\cdot3$

23

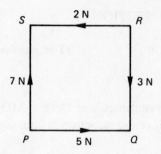

Forces act along the sides of a square $PQRS$ as shown above. The resultant of the system

1 has magnitude 5 N   **2** acts at an angle $\tan^{-1}(4/3)$ to PQ

3 passes through P

24 The displacement, x m, of a particle at time t seconds is given by

$$x = 4\cos 2t - 3\sin 2t.$$

1 When $t = \pi/2$ the speed is 8 m s^{-1}

2 The maximum speed is 10 m s^{-1}

3 The motion is simple harmonic

25 Two smooth elastic spheres of equal radii, but of unequal masses and moving with unequal velocities, collide directly. The coefficient of restitution between the spheres is e, where $0 < e < 1$.

1 The total kinetic energy of the two spheres is unaltered by the collision

2 The velocity of each sphere after the collision is its velocity before the collision multiplied by $-e$

3 The combined momentum of the two spheres is unaltered by the collision

26 A particle P, attached to a fixed point by a light inextensible string, is moving round a complete circle in a vertical plane.

1 The sum of the kinetic and potential energies of P is constant

2 When the velocity of P is horizontal, its acceleration is vertical

3 When the velocity of P is vertical, its acceleration is horizontal

27 A ball is thown at an angle of $30°$ to a horizontal plane with speed 40 m s^{-1}.

1 It reaches its maximum height after 2 s

2 The maximum height reached is 20 m

3 The range on the plane is 80 m

28 A parcel of mass m lies on horizontal ground.

1 The weight of the parcel is mg

2 The force exerted by the ground on the parcel is $-mg$

3 The force exerted by the parcel on the ground is mg

29 A constant force of magnitude 6 N acts on a body of mass 12 kg and is the only force acting on the body. At a particular instant, the body has a speed of 3 m s^{-1}. At this instant

1 the acceleration of the body is 2 m s^{-2}

2 the kinetic energy of the body is 54 J

3 the power developed by the force is 18 W

30

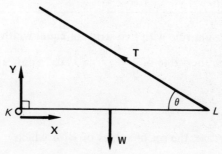

A uniform rod KL, of weight $\mathbf{W}$, is hinged to a smooth pivot at K and is supported in a horizontal position by a string attached at L. For the forces shown

1 $|\mathbf{X}| = |\mathbf{T}| \sin \theta$ **2** $|\mathbf{Y}| = |\mathbf{W}| - |\mathbf{T}| \cos \theta$

3 $2|\mathbf{T}| \sin \theta - |\mathbf{W}| = 0$

6   Longer tests: pure mathematics

6.1   Paper 5

Time allowed: $2\frac{1}{2}$ hours

Answer as many questions as you can.

1   Prove, by induction or otherwise, that

$$\sum_{r=1}^{n} \frac{1}{(r+1)(r+2)} = \frac{n}{2(n+2)}.$$

(5 marks)

---

2   Find $\dfrac{dy}{dx}$ in terms of x when

(a) $y = \ln\left(\dfrac{1+x^2}{1-x^2}\right)$,   (b) $y = e^{-2x}\sin(3x)$.

(6 marks)

---

3   Use the trapezium rule with five strips of equal width to estimate, to three decimal places, the value of $\displaystyle\int_0^{\pi/6} \sqrt{(\sin x)}\, dx$. Show all your working.

(6 marks)

---

4   Find, in each case, the set of values of x for which

(a) $|x - 1| < |2x - 1|$,   (b) $|\cos x| \geqslant \frac{1}{2}$.

(7 marks)

5   Solve the differential equation

$$xy\frac{dy}{dx} = 1 + y^2$$

given that $y = 1$ when $x = 1$. Hence express y in terms of x.

(7 marks)

---

6   Given that $z = x + iy$, where $x, y \in \mathbb{R}$, and $z^2 = 5 - 12i$, find the two possible values of z. Show, in an Argand diagram, the points A and B representing these roots.

(8 marks)

---

7   Solve, for $0 \leqslant x \leqslant 2\pi$, the equations

(a) $\sin x = -\frac{1}{2}$,   (b) $3 \cos 2x + 7 \cos x = 0$,

giving your answers to (b) to two decimal places.

(8 marks)

---

8   Find equations of the tangent and the normal at the point $P(at^2, 2at)$ of the parabola $y^2 = 4ax$, where $a > 0$, $t > 0$.

Given that the focus S of the parabola is the point $(a, 0)$ and the tangent and normal at P meet the x-axis at T and G respectively, prove that TP and GP bisect the angles between SP and the line through P which is parallel to the x-axis.

(9 marks)

---

9   Referred to the origin O, the position vectors of the points A, B are $\mathbf{a}, \mathbf{b}$ respectively. Show that the vector equation of the line AB can be written in the form

$$\mathbf{r} = \mathbf{a} + t(\mathbf{b} - \mathbf{a}).$$

The points C, D have position vectors $\frac{1}{3}\mathbf{a}, \frac{2}{3}\mathbf{b}$ respectively. Find the position vector of the point of intersection Q of BC and AD.

(9 marks)

---

10 Given that
$$f(x) \equiv x^3 + 3x + 2,$$
show that $f'(x) > 0$ for all $x \in \mathbb{R}$.

Deduce that the equation $f(x) = 0$ has just one real root, which lies between -1 and 0. Use an iterative process to estimate this root to three decimal places. Sketch the graph of $y = f(x)$.

(11 marks)

---

11 Evaluate the integrals

(a) $\displaystyle\int_0^1 x \cos(\pi x)\, dx,$ (b) $\displaystyle\int_0^3 x\sqrt{(x+1)}\, dx,$ (c) $\displaystyle\int_0^{\pi/4} \tan^2 x\, dx.$

(11 marks)

---

12 Express $f(x)$, where
$$f(x) \equiv \frac{x-3}{(x^2+3)(x+1)},$$
in partial fractions.

(a) Show that $\displaystyle\int_3^4 f(x)\, dx = \tfrac{1}{2}\ln(\tfrac{76}{75}).$

(b) Find the coefficient of x^{2n+1} in the expansion of $f(x)$ in ascending powers of x.

(12 marks)

---

13 Find the values of R and ϕ when $(8\cos x - 15\sin x)$ is expressed in the form $R\cos(x+\phi)$, where R is positive and ϕ is an acute angle, giving the value of ϕ to the nearest tenth of a degree.

Hence

(a) find the maximum and minimum values of the expression
$$\frac{5}{18 + 8\cos x - 15\sin x},$$

(b) find, to the nearest degree, the values of x between $0°$ and $360°$ which satisfy the equation
$$8\cos x - 15\sin x = 10.$$

(12 marks)

---

14 Functions f, g, h are defined by

$$f(x) = e^{-x}, \qquad x \in \mathbb{R},$$

$$g(x) = \frac{x}{1 + x^2}, \qquad x \in \mathbb{R},$$

$$h(x) = \ln \sec x, \qquad -\pi/4 < x < \pi/4.$$

(a) Determine whether the functions are odd, even or neither.

(b) For each function find the range.

(c) For each function, determine whether an inverse function exists.

If the inverse function exists, give an expression for it; if not, suggest a suitable subdomain so that the function, restricted to this subdomain, would have an inverse.

(14 marks)

6.2 Solutions and mark scheme for Paper 5

(Note the acceptable abbreviations used in the solutions.)

1 Assume that

$$\sum_{r=1}^{n} \frac{1}{(r + 1)(r + 2)} = \frac{n}{2(n + 2)}. \tag{1}$$

Then

$$\sum_{r=1}^{n+1} \frac{1}{(r + 1)(r + 2)} = \sum_{r=1}^{n} \frac{1}{(r + 1)(r + 2)} + \frac{1}{(n + 1 + 1)(n + 1 + 2)}$$

$$= \frac{n}{2(n + 2)} + \frac{1}{(n + 2)(n + 3)}$$

$$= \frac{n(n + 3) + 2}{2(n + 2)(n + 3)}$$

$$= \frac{n^2 + 3n + 2}{2(n + 2)(n + 3)} = \frac{(n + 1)(n + 2)}{2(n + 2)(n + 3)}$$

$$\Rightarrow \sum_{r=1}^{n+1} \frac{1}{(r + 1)(r + 2)} = \frac{(n + 1)}{2(n + 3)} = \frac{n + 1}{2(n + 1 + 2)}. \tag{M1 A2}$$

But this is the same formula as (1) except that n is replaced by $(n + 1)$. Therefore, if result (1) is true for n, it is also true for $n + 1$.

But when $n = 1$, left-hand side $= \dfrac{1}{2 \cdot 3} = $ right-hand side. **(M1)**

Therefore the result is true for $n = 1$.

Therefore the result is true for $1 + 1 = 2$, $2 + 1 = 3, \ldots$, and so on universally. **(Completion A1 cao)**

2 (a) $\quad y = \ln(1 + x^2) - \ln(1 - x^2)$ **(Splitting log M1)**

$\Rightarrow \dfrac{dy}{dx} = \dfrac{2x}{1 + x^2} - \dfrac{(-2x)}{1 - x^2}$ **(Differentiation of log M1)**

$\qquad = 2x\left(\dfrac{1}{1 + x^2} + \dfrac{1}{1 - x^2}\right).$ **(A1 cao)**

(b) $\dfrac{dy}{dx} = e^{-2x}\, 3\cos(3x) + (-2)\, e^{-2x} \sin(3x)$ **(M1 A1 cao, A1 cao)**

$\qquad = e^{-2x}[3\cos(3x) - 2\sin(3x)].$

3

x	0	6°	12°	18°	24°	30°	
$\sin x$	0	0·1045	0·2079	0·3090	0·4067	0·5	**(Table 2)**
$\sqrt{(\sin x)}$	0	0·3233	0·4560	0·5559	0·6378	0·7071	

$\displaystyle\int_0^{\pi/6} \sqrt{(\sin x)}\, dx \approx \dfrac{1}{2}\dfrac{\pi}{30}\big[0 + 2(0\!\cdot\!3233 + 0\!\cdot\!4560 + 0\!\cdot\!5559$

$\qquad\qquad\qquad\qquad\qquad + 0\!\cdot\!6378) + 0\!\cdot\!7071\big]$

$\qquad\qquad = \dfrac{\pi}{60}\big[0\!\cdot\!7071 + 2 \times 1\!\cdot\!9730\big]$

(Correct use of correct formula M1)

$\qquad\qquad = \dfrac{\pi}{60} \times 4\!\cdot\!6531 \approx 0\!\cdot\!2436$ **(A2)**

$\Rightarrow \displaystyle\int_0^{\pi/6} \sqrt{(\sin x)}\, dx \approx 0\!\cdot\!244$ to three decimal places.

(A1 cao)

---

4 (a) $\qquad\qquad\qquad (x - 1)^2 < (2x - 1)^2$ **(M1)**

$\Rightarrow \quad (2x - 1)^2 - (x - 1)^2 > 0$

$\Rightarrow \qquad\qquad x(3x - 2) > 0$ **(M1 A1 cao)**

$\Rightarrow \qquad\qquad x < 0 \quad$ or $\quad x > \tfrac{2}{3}$

(A1 cao)

or $\qquad\qquad \{x : x < 0\} \cup \{x : x > \tfrac{2}{3}\}.$

(b) $\quad \cos x = \pm\tfrac{1}{2} \qquad$ when $x = n\pi \pm \pi/3,\, n \in \mathbb{Z}$

$\Rightarrow \quad |\cos x| \geqslant \tfrac{1}{2} \qquad$ when $n\pi - \pi/3 \leqslant x \leqslant n\pi + \pi/3,\, n \in \mathbb{Z}.$

(M1 A1 A1)

---

5 Separating variables

$$\frac{y}{1+y^2}\,dy = \frac{1}{x}\,dx \qquad \textbf{(M1)}$$

$$\Rightarrow \quad \int\frac{y}{1+y^2}\,dy = \int\frac{1}{x}\,dx + C \qquad \textbf{(Constant B1)}$$

$$\Rightarrow \quad \tfrac{1}{2}\ln(1+y^2) = \ln x + C \qquad \textbf{(Integrations A1 A1)}$$

$$\Rightarrow \quad \ln(1+y^2) = \ln(Ax^2)$$

$$\Rightarrow \quad 1+y^2 = Ax^2. \qquad \textbf{(Clearing logs M1)}$$

$$y = 1 \quad \text{when } x = 1 \quad \Rightarrow \quad A = 2 \quad \textbf{(M1)}$$

$$\Rightarrow \quad y^2 = 2x^2 - 1$$

$$\Rightarrow \quad y = \sqrt{(2x^2 - 1)}, \qquad \textbf{(A1 cao)}$$

taking the $+$ve root since $y = 1$ when $x = 1$.

6

$$(x + iy)^2 = 5 - 12i \qquad (1)$$

$$\Rightarrow \quad x^2 - y^2 = 5 \quad \text{(equating real parts).} \qquad (2) \qquad \textbf{(M1)}$$

Also changing i into $-i$ in (1)

$$\Rightarrow \quad (x - iy)^2 = 5 + 12i. \qquad (3) \qquad \textbf{(M1)}$$

$$(1) \times (3) \quad \Rightarrow \quad (x^2 + y^2)^2 = 5^2 + 12^2 = 169$$

$$\Rightarrow \quad x^2 + y^2 = 13, \qquad (4)$$

discarding the $-$ve sign in the $\sqrt{}$ since $x^2 + y^2 > 0$. $\qquad$ **(A1)**

$$(2) + (4) \quad \Rightarrow \quad 2x^2 = 18 \quad \Rightarrow \quad x^2 = 9 \quad \Rightarrow \quad x = \pm 3,$$
$$(4) - (2) \quad \Rightarrow \quad 2y^2 = 8 \quad \Rightarrow \quad y^2 = 4 \quad \Rightarrow \quad y = \pm 2.$$
(M1 A1 both)

But equating imaginary parts in (1) $\Rightarrow 2xy = -12$ and so x, y have opposite signs. $\qquad$ **(M1)**

Therefore

$$x = 3, y = -2 \qquad \text{and} \qquad x = -3, y = 2$$

$$\Rightarrow \quad z = \pm (3 - 2i). \qquad \textbf{(A1 FT)}$$

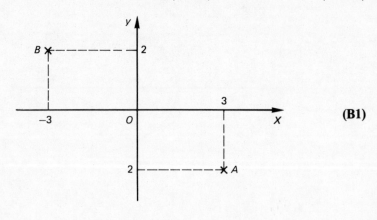

(B1)

---

7 (a) $\qquad \sin x = -\frac{1}{2} \quad \Rightarrow \quad \sin x = \sin(\pi + \pi/6) = \sin(2\pi - \pi/6)$
$\qquad\qquad\qquad\qquad\qquad\qquad\qquad\qquad\qquad$ **(M1 A1)**

$$\Rightarrow \quad x = 7\pi/6 \quad \text{or} \quad 11\pi/6. \qquad \textbf{(Both roots A1 cao)}$$

(b) $\qquad\qquad\qquad 3(2\cos^2 x - 1) + 7\cos x = 0 \qquad \textbf{(M1)}$

$$\Rightarrow \quad 6\cos^2 x + 7\cos x - 3 = 0$$

$$\Rightarrow \quad (3\cos x - 1)(2\cos x + 3) = 0 \qquad \textbf{(M1)}$$

$$\Rightarrow \quad \cos x = \tfrac{1}{3} \quad \text{or} \quad \cos x = -\tfrac{3}{2}. \qquad \textbf{(A1)}$$

$$\cos x = \tfrac{1}{3} \quad \Rightarrow \quad x \approx 1.23 \quad \text{or} \quad 5.05.$$

(**Both answers A1 cao**)

Since $|\cos x| \leqslant 1$ for $x \in \mathbb{R}$, there is no real value of x for which $\cos x = -\tfrac{3}{2}$. **(M1)**

8

$$\frac{\mathrm{d}y}{\mathrm{d}x} = \frac{\dot{y}}{\dot{x}} = \frac{2a}{2at} = \frac{1}{t}. \qquad \textbf{(M1)}$$

Tangent is

$$(y - 2at) = \frac{1}{t}(x - at^2) \equiv x - ty + at^2 = 0. \qquad \textbf{(A1 cao)}$$

Normal is

$$(y - 2at) = -t(x - at^2) \equiv tx + y - 2at - at^3 = 0$$

(M1 A1 cao)

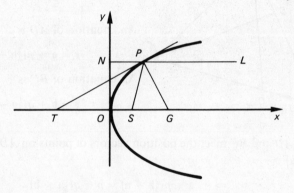

$$T \text{ is } (-at^2, 0) \quad \Rightarrow \quad ST = a + at^2.$$

$$SP^2 = (at^2 - a)^2 + (2at)^2 = a^2(t^4 + 2t^2 + 1) = \left[a(t^2 + 1)\right]^2.$$

Therefore $SP = ST \Rightarrow S\hat{P}T = S\hat{T}P$. **(M1 A2)**
But $S\hat{T}P = T\hat{P}N$ ($ST \parallel PN$).
Therefore $S\hat{P}T = T\hat{P}N$.
Also $S\hat{P}G = 90° - S\hat{P}T$, $G\hat{P}L = 180° - 90° - T\hat{P}N = 90° - S\hat{P}T$.
Therefore $S\hat{P}G = G\hat{P}L$. **(M1 A1)**

9

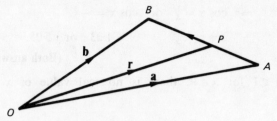

By the triangle of vectors $\overrightarrow{OA} + \overrightarrow{AB} = \overrightarrow{OB}$

$$\Rightarrow \quad \overrightarrow{AB} = \mathbf{b} - \mathbf{a}$$

$$\Rightarrow \quad \overrightarrow{AP} = t(\mathbf{b} - \mathbf{a}),$$

where t is a parameter $(= AP/AB)$.

Then $\qquad \mathbf{r} = \overrightarrow{OP} = \overrightarrow{OA} + \overrightarrow{AP} = \mathbf{a} + t(\mathbf{b} - \mathbf{a}).$ **(M2 A1)**

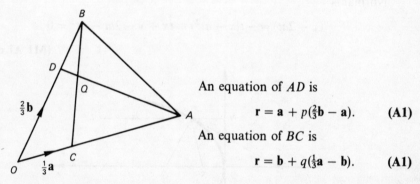

An equation of AD is

$$\mathbf{r} = \mathbf{a} + p(\tfrac{2}{3}\mathbf{b} - \mathbf{a}). \qquad \textbf{(A1)}$$

An equation of BC is

$$\mathbf{r} = \mathbf{b} + q(\tfrac{1}{3}\mathbf{a} - \mathbf{b}). \qquad \textbf{(A1)}$$

Where AD and BC meet the position vectors of points on AD and BC must be equal

$$\Rightarrow \quad \mathbf{a} + p(\tfrac{2}{3}\mathbf{b} - \mathbf{a}) = \mathbf{b} + q(\tfrac{1}{3}\mathbf{a} - \mathbf{b})$$

$$\Leftrightarrow \quad (1-p)\mathbf{a} + \tfrac{2}{3}p\mathbf{b} = \tfrac{1}{3}q\mathbf{a} + (1-q)\mathbf{b}.$$

Equating coefficients of $\mathbf{a}$ and $\mathbf{b}$,

$$\Rightarrow \quad 1 - p = \tfrac{1}{3}q, \qquad \tfrac{2}{3}p = 1 - q$$

$$\Leftrightarrow \qquad p = \tfrac{6}{7}, \qquad q = \tfrac{3}{7}. \qquad \textbf{(M2 A1)}$$

The position vector of Q is $(\mathbf{a} + 4\mathbf{b})/7$. **(A1 cao)**

---

10 $\qquad f'(x) = 3x^2 + 3 > 0 \quad \forall x \in \mathbb{R} \quad \text{since } x^2 \geqslant 0 \text{ and } 3 > 0.$ **(2)**

$f'(x) > 0 \Rightarrow f(x)$ is an increasing function $\forall x$ and so can pass through any stated value only once

$$\Rightarrow \quad f(x) = 0 \quad \text{for one value of } x \text{ only.} \qquad \textbf{(A1)}$$

$f(-1) = -2$, $f(0) = 2$ and so $f(x) = 0$ for one value of x in $(-1, 0)$.

(M1 A1)

Let us take a first approximation $x_1 = -0.5$ in the Newton–Raphson formula

$$x_{n+1} = x_n - \frac{f(x_n)}{f'(x_n)}.$$ **(M1)**

$$x_2 = -0.5 - \frac{(-0.125 - 1.5 + 2)}{0.75 + 3}$$

$$= -0.5 - \frac{0.375}{3.75} = -0.6.$$ **(A1)**

$$x_3 = -0.6 - \frac{(-0.216 - 1.8 + 2)}{1.08 + 3}$$

$$= -0.6 + \frac{0.016}{4.08} \approx -0.6 + 0.0039 = -0.596(1).$$ **(A1)**

$f(-0.5961) = -0.211\ 815\ 3 - 1.7883 + 2 \approx -0.000\ 115.$

The next correction would be approximately $-0.000\ 03$ and so the root is -0.596 to three decimal places. **(Check A1)**

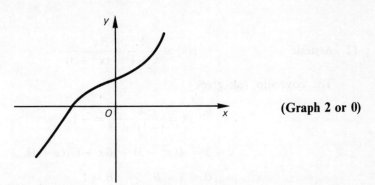

(Graph 2 or 0)

---

11 (a) $\int x \cos (\pi x)\, dx = \int x \frac{d}{dx}\left[\frac{1}{\pi} \sin (\pi x)\right] dx$ **(M1)**

$$= \frac{x}{\pi} \sin (\pi x) - \int \frac{1}{\pi} \sin (\pi x)\, dx$$ **(A1)**

$$= \frac{x}{\pi} \sin (\pi x) + \frac{1}{\pi^2} \cos (\pi x)$$ **(A1 cao)**

$$\Rightarrow \int_0^1 x \cos (\pi x)\, dx = \frac{1}{\pi} \sin \pi + \frac{1}{\pi^2} \cos \pi - 0 - \frac{1}{\pi^2} \cos 0$$

$$= -2/\pi^2. \qquad \text{(A1 cao)}$$

(b)
$$1 + x = t^2 \quad \Rightarrow \quad dx = 2t\, dt. \qquad \text{(Substitution M1)}$$

$$x = 0 \quad \Rightarrow \quad t = 1, \qquad x = 3 \quad \Rightarrow \quad t = 2. \qquad \text{(Limits A1)}$$

$$I = \int_1^2 (t^2 - 1)t \cdot 2t\, dt = 2 \int_1^2 (t^4 - t^2)\, dt \qquad \text{(A1 cao)}$$

$$= 2 \left[\frac{t^5}{5} - \frac{t^3}{3} \right]_1^2 = 2 \left(\frac{32}{5} - \frac{8}{3} - \frac{1}{5} + \frac{1}{3} \right) = \frac{116}{15}. \qquad \text{(A1 cao)}$$

(c)
$$\int_0^{\pi/4} \tan^2 x\, dx = \int_0^{\pi/4} (\sec^2 x - 1)\, dx \qquad \text{(M1)}$$

$$= \left[\tan x - x \right]_0^{\pi/4} \qquad \text{(A1)}$$

$$= 1 - \pi/4. \qquad \text{(A1 cao)}$$

---

12 Assume
$$f(x) \equiv \frac{A}{x + 1} + \frac{Bx + C}{(x^2 + 3)}. \qquad \text{(M1)}$$

The 'cover-up' rule gives

$$A = \frac{-1 - 3}{(-1)^2 + 3} = -1. \qquad \text{(A1)}$$

$$x - 3 = A(x^2 + 3) + (Bx + C)(x + 1). \qquad (1)$$

$$x^2 \quad \Rightarrow \quad 0 = A + B \quad \Rightarrow \quad B = 1.$$
$$\qquad \qquad \qquad \qquad \qquad \qquad \text{(M1 A1 both cao)}$$
$$x^0 \quad \Rightarrow \quad -3 = 3A + C \quad \Rightarrow \quad C = 0.$$

$$f(x) \equiv \frac{-1}{x + 1} + \frac{x}{x^2 + 3}.$$

(a)
$$\left[-\ln(x + 1) + \tfrac{1}{2} \ln(x^2 + 3) \right]_3^4 \qquad \text{(M1 logs, A2, 1 or 0 FT)}$$

$$= -\ln 5 + \tfrac{1}{2} \ln 19 + \ln 4 - \tfrac{1}{2} \ln 12$$

$$= \tfrac{1}{2} \ln(\tfrac{19}{12}) + \ln(\tfrac{4}{5}) = \tfrac{1}{2} \ln \left(\frac{19 \times 4^2}{12 \times 5^2} \right) = \tfrac{1}{2} \ln(\tfrac{76}{75}).$$

$$\text{(Completion A1 cao)}$$

(b) Coefficient of x^{2n+1} in $-(1+x)^{-1}$ is

$$-(-1)^{2n+1} = 1. \quad \text{(A1 FT)}$$

Coefficient of x^{2n+1} in $\dfrac{x}{(x^2+3)}\left[\equiv \dfrac{x}{3}\left(1+\dfrac{x^2}{3}\right)^{-1}\right]$ is

$$\frac{1}{3}(-1)^n\frac{1}{3^n} = (-1)^n\frac{1}{3^{n+1}}. \quad \text{(M1 A1 FT)}$$

The required coefficient is $1 + \dfrac{(-1)^n}{3^{n+1}}$. \hfill (A1 cao)

13

$$R\cos(x+\phi) \equiv R\cos x\cos\phi - R\sin x\sin\phi$$

$$\Rightarrow \quad R\cos x\cos\phi - R\sin x\sin\phi \equiv 8\cos x - 15\sin x$$

$$\Rightarrow \quad R\cos\phi = 8, \qquad R\sin\phi = 15 \quad \text{(M2 A1)}$$

$$\Rightarrow \quad R^2 = 8^2 + 15^2 = 17^2, \qquad \tan\phi = \tfrac{15}{8} = 1\cdot875$$

$$\Rightarrow \quad R = 17, \quad \text{(A1)}$$

$$\phi \approx 61\cdot92° \text{ or } 61\cdot9° \text{ (to the nearest tenth of a degree).} \quad \text{(A1)}$$

(a)
$$\frac{5}{18+8\cos x - 15\sin x} = \frac{5}{18 + 17\cos(x+\phi)}.$$

The greatest and least values of the numerator occur when $\cos(x+\phi) = \pm1$ \hfill (M1)

$$\Rightarrow \quad \text{least value} = \frac{5}{18+17} = \frac{1}{7}, \quad \text{(A1 cao)}$$

$$\text{greatest value} = \frac{5}{18-17} = 5. \quad \text{(A1 cao)}$$

(b) The given equation can be written

$$17\cos(x+\phi) = 10$$

$$\Rightarrow \quad \cos(x+\phi) = \tfrac{10}{17} \approx \cos 53\cdot97°$$

$$\Rightarrow \quad x + 61\cdot92° \approx (360n \pm 53\cdot97)°. \quad \text{(M1 A1)}$$

The two values of x in the range $0°$ to $360°$ are $(360 - 53\cdot97 - 61\cdot92)°$ and $(360 + 53\cdot97 - 61\cdot92)°$, i.e. $244°$ and $352°$ to the nearest degree.

\hfill (A1 A1 cao)

---

14 (a) $g(x)$ is odd, $h(x)$ is even, $f(x)$ is neither.

(A2, 1 or 0, one off for each error)

(b) As x increases from $-\infty$ to ∞, $f(x)$ decreases (steadily) from ∞ to 0 and so the range of $f(x)$ is $\mathbb{R}^+$. **(A1 cao)**

$$g'(x) = \frac{(1 + x^2) - 2x^2}{(1 + x^2)^2} = \frac{1 - x^2}{(1 + x^2)^2} = \frac{(1 - x)(1 + x)}{(1 + x^2)^2}$$

$\Rightarrow \quad g(x)$ has a maximum $(\frac{1}{2})$ when $x = 1$

and, because it is an odd function, a minimum $(-\frac{1}{2})$ when $x = -1$ **(M1 A1)**

$\Rightarrow \quad$ range is $-\frac{1}{2} \leqslant g(x) \leqslant \frac{1}{2}$ or $[-\frac{1}{2}, \frac{1}{2}]$. **(A1 cao)**

$h(x)$ is even and $\sec x$ increases from 1 to $\sqrt{2}$ as x increases from 0 to $\pi/4$, so the range is

$$0 \leqslant h(x) \leqslant \tfrac{1}{2} \ln 2 \quad or \quad [0, \tfrac{1}{2} \ln 2].$$ **(A1 cao)**

(c) $f(x) = e^{-x}$ is a one:one mapping and the inverse is defined by $x = e^{-y}$

$\Rightarrow \qquad y = -\ln x$

$\Rightarrow \quad f^{-1}(x) = -\ln x \quad$ for $x \in \mathbb{R}^+$. **(M1 A1)**

Sketches of the graphs of g and h show that inverse functions do not exist for the domains specified. **(M1)**

The restricted domains $-1 \leqslant x \leqslant 1$ for g **(M1 A1)** and $0 \leqslant x \leqslant \pi/4$ for h **(M1 A1)** make g and h one:one functions, for which an inverse function exists in each case.

6.3 Explanation of solutions and mark scheme for Paper 5

1 The essential point of induction is to *assume* that the given formula is true for n, and use this to prove that the same formula is true with $n + 1$ in place of n. Then the truth of the formula for the lowest possible value of n, in this case 1, should be independently proved (usually by substitution). Then the final statement completes the proof.

Note that the question states 'Prove by induction, or otherwise . . . .' The 'otherwise' is as follows (using the method of differences):

By partial fractions

$$\frac{1}{(r+1)(r+2)} \equiv \frac{1}{r+1} - \frac{1}{r+2} \qquad \textbf{(M1 A1)}$$

$$\Rightarrow \sum_{r=1}^{n} \frac{1}{(r+1)(r+2)} = \frac{1}{2} - \frac{1}{3}$$

$$+ \frac{1}{3} - \frac{1}{4}$$

$$+ \ldots$$

$$\ldots$$

$$+ \frac{1}{n+1} - \frac{1}{n+2}$$

$$= \frac{1}{2} - \frac{1}{n+2} \qquad \textbf{(M1 A1)}$$

$$= \frac{n}{2(n+2)}. \qquad \textbf{(A1 cao)}$$

---

2 (a) Note that we use the relation $\ln\left[f(x)/g(x)\right] = \ln f(x) - \ln g(x)$ to simplify the work and **(M1)** is given for realizing this. Then we use $\dfrac{d}{dx}\left[\ln f(x)\right] = \dfrac{f'(x)}{f(x)}$ **(M1)** to differentiate the terms (both correct for **A1**).

Always try and simplify your working. You could have tried

$$\frac{1}{(1+x^2)/(1-x^2)} \frac{d}{dx}\left(\frac{1+x^2}{1-x^2}\right),$$

but you are more likely to make a mistake, with **(M1)** for knowing how to differentiate a logarithm, **(M1)** for differentiating a quotient and **(A1 cao)** for the final result.

(b) Here you receive **(M1)** for trying to differentiate a product and **(A1 cao)** for each correct term.

---

3 We tabulate (using degrees) *by calculator*. Values of $\sin x$ are not strictly necessary and could be omitted, but the question states 'show all your working'.

We gain **(M1)** for attempting to use the correct trapezium formula for integration, $\frac{1}{2}h\left[F + L + 2 \times \text{rest}\right]$. We retain four places of decimals only— you could keep five places to be sure but, in approximate integration with small h, retention of one extra decimal place is usually sufficient.

Check that the answer seems reasonable thus: The average value of $\sqrt{(\sin x)}$ for $0 \leqslant x \leqslant \pi/6$ is approximately 0·5 and the range of integration $\pi/6 \approx 0\cdot5$, suggesting an answer of approximately 0·25.

---

4 (a) The best way to deal with the inequation $|f(x)| > |g(x)|$ is to write it in the equivalent form

$$[f(x)]^2 > [g(x)]^2 \quad \Rightarrow \quad [f(x)]^2 - [g(x)]^2 > 0$$

and then transform to (using the difference of two squares)

$$[f(x) - g(x)][f(x) + g(x)] > 0.$$

Alternatively, you could consider the cases

$$x < -\tfrac{1}{2} \quad \Rightarrow \quad -(x-1) < -(2x+1),$$
$$-\tfrac{1}{2} \leqslant x \leqslant 1 \quad \Rightarrow \quad -(x-1) < 2x+1,$$
$$1 < x \quad \Rightarrow \quad x - 1 < 2x + 1,$$

and combine the various results.

(b) Note the general solution is given and that the signs are $\geqslant$, $\leqslant$. The graph of $\cos x$ for $0 \leqslant x \leqslant 2\pi$ is useful here. Note also that the answer is given in radians.

---

5 Note the **(B1)** mark for introducing the arbitrary constant at some stage. The next two **(A)** marks for integration would be awarded even if a constant was not introduced, but would, of course, be dependent on the **(M1)** for separating variables. Also the **(M1)** for clearing logs would be independent of the **(B1)** for the arbitrary constant but the last two marks cannot be earned without the **(B1)**.

Many candidates miss the last **(A1 cao)** mark—either they tend to write $y = \pm\sqrt{(2x - 1)}$, and do not discard the negative sign, or give no reason for choosing the positive sign (only).

---

6 The given solution is self-explanatory. Remember that *any* relation between complex numbers remains true when you change i into $-$i and this simplifies the working in this case.

You could have equated real and imaginary parts in (1) to obtain

$$x^2 - y^2 = 5, \qquad 2xy = -12 \qquad \textbf{(M1 A1 A1)}$$

and then solved these equations **(M1 A1)** and chosen the correct signs **(M1)**. Note the last **(B)** mark, which is independent of your having earned the M marks, but is given only if you accurately show your roots on your sketch of the Argand diagram. Squared paper is advisable here.

---

7 Note that, in this question, the $0 \leqslant x \leqslant 2\pi$ means that you must find the solution in radians.

(a) You should be able to obtain the answers from $\sin^{-1}\left(\frac{1}{2}\right) = \pi/6$ and looking at the graph of $\sin x$.

(b) The only satisfactory way to solve this problem is to obtain an equation in $\cos x$. Solve this quadratic (by factorization) and hence find the values of x (radians *not* degrees) using your calculator.

The examiners usually expect some reason (to be given) why there is no real solution to the equation $\cos x = -\frac{3}{2}$.

---

8 Marks may well be given for the unreduced forms of the tangent and normal (i.e. those with brackets etc.) but it is always useful to simplify the equations and check that they have the correct gradients and pass through P.

Here we have used pure geometry by using the fact that proving $S\hat{P}T = T\hat{P}N$ is equivalent to proving $ST = SP$. Of course you can complete this question by calculating $\tan P\hat{T}S$ and $\tan S\hat{P}T$ and showing that they are equal, but this is more complicated.

Try and use simple geometrical properties of curves, especially circles, whenever possible.

---

9 The position vector of Q is found by obtaining the position vector of an arbitrary point on each of the lines AD and BC. Where these lines intersect, these position vectors must be equal. We then use the result that, if $\mathbf{a}, \mathbf{b}$ are non-zero, non-parallel vectors and

$$\lambda \mathbf{a} + \mu \mathbf{b} = \lambda_1 \mathbf{a} + \mu_1 \mathbf{b},$$

then $\lambda = \lambda_1$, $\mu = \mu_1$.

The final result, for the position vector of Q, can be checked by substitution in the equations of the lines AD and BC.

---

10 Notice that we give reasons for the three facts to be proved, i.e. for $f'(x) > 0$, that the equation $f(x) = 0$ has only one root and that this root lies in $(-1, 0)$. Agreement with the examiner but no proof receives no credit!

Calculators must be used in our iterative process, the working being shown, and a check on the third decimal place is required for full marks. The sketch needs only to be simple. The word 'sketch' means a general indication of the shape and position of the curve, i.e. where it crosses the axes and what happens when $|x|$ is large.

---

11 (a) The method mark is given for appreciating that integration by parts is needed *and going the correct way about it.*

After evaluation of the indefinite integral you can easily check your accuracy by differentiation.

Do not be alarmed that the answer is negative. Look at the areas between the curve $y = x \cos(\pi x)$ and Ox and you will find that the area below Ox ($-$ve) is greater than the area above Ox ($+$ve).

(b) The substitution has been chosen so as to remove the square root.

It is not necessary to alter the limits. You can evaluate the indefinite integral for x and then put in the limits for x.

(c) Usually this type of integral requires the use of some trigonometrical identity, thereby transforming the integral into a standard form.

---

12 If you fail to make the correct assumption for the form of the partial fractions, you get no marks at all for the partial fractions.

If preferred, you could use equation (1), comparing the coefficients of x^1 $[1 = B + C]$, to obtain a third equation between A, B, C and so avoid the 'cover-up' rule.

Note that the correct expression for $f(x)$ is written out in full—it got no mark as such, but is useful for reference and as an easy check for correctness. Always make a quick check in these problems.

(a) **(M1)** is given for appreciating that the integration gives rise to logs.

The final **(A1)** is for combining logs, using $-\log x = \log(1/x)$ and cleaning up correctly.

(b) We have used the binomial expansion in the form

$$(1 + X)^{-1} = 1 - X + X^2 - \ldots + (-1)^r X^r + \ldots .$$

Before expanding $(x^2 + 3)^{-1}$, it must be expressed as $\dfrac{1}{3}\left(1 + \dfrac{x^2}{3}\right)^{-1}$ and then the above expansion for $(1 + X)^{-1}$ with $X = x^2/3$ can be used.

Note that incorrect partial fractions are followed through in (a) and (b) where possible. However, a printed answer, as in (a), will always be **cao**.

---

13 The equations of line 3 are derived by equating the coefficients of $\cos x, \sin x$ on the two sides of the equation of line 2.

(a) Do not use calculus here. The question says 'hence' and it means just this.

Remember that $\cos(x + \phi)$ lies between -1 and 1 and that $\dfrac{1}{f(x)}$ has its greatest and least values when $f(x)$ has its least and greatest values respectively.

(b) Remember that any equation of the form

$$\left.\begin{array}{c}\cos\\\sin\end{array}\right\}(x + \phi) = k,$$

where $|k| < 1$, has two roots in any interval of width $360°$ (or 2π).

Although we have used a calculator in the numerical part of this question, we correct the answers as required. However, with a calculator it is easy to carry two extra significant figures as a guard against rounding errors.

---

14 Throughout this question, we are assuming some understanding of the behaviour of f, g as $x \to \pm\infty$ and of the graphs of all three functions.

(a) Remember that the function $k(x)$ is
   (i) even if $k(-x) = k(x)$,
   (ii) odd if $k(-x) = -k(x)$.
(b) Note that, to find the range of $g(x)$, it is necessary to find its maximum value. We also use the fact that $g(x)$ is an odd function.
(c) The inverse of $g(x)$ could be found by solving the equation $x = g(y)$ for y in terms of x, i.e. interchanging x and y.

   The existence of an inverse for a function $k(x)$ requires a one:one mapping of the inverse. Subdomains have been chosen accordingly.

6.4 Paper 6

Time allowed: $2\frac{1}{2}$ hours

Answer as many questions as you can.

1 Referred to the origin O, the points A, B, C have position vectors

$$2\mathbf{i} + \mathbf{j} - \mathbf{k}, \qquad \mathbf{i} + 3\mathbf{j} + 3\mathbf{k}, \qquad -\mathbf{i} + \mathbf{j} + 5\mathbf{k}$$

respectively. Find the cosine of the angle BAC.
   Find also the area of this triangle.

(6 marks)

2 Find the coordinates of the centres A, B of the circles

$$x^2 + y^2 - 6x + 8y - 9 = 0,$$
$$x^2 + y^2 + 4x - 16y - 13 = 0,$$

respectively.

Prove that these circles touch externally and find an equation of the circle on AB as diameter.

(6 marks)

---

3 Find the number of permutations of the seven letters of the word ADDRESS

(a) in which the letter R is the middle letter,

(b) in which the letter R does not come first or last.

(6 marks)

---

4 Find the value, when $x = 1$, of the derivative with respect to x of

(a) $\dfrac{x}{1 + 2x^3}$, (b) $\tan^{-1}(x/3)$.

(6 marks)

---

5 Find the complete set of values of x for which

$$0 < \frac{x(2 + x)}{2x + 1} < 1.$$

(7 marks)

---

6 Find, in radians, the general solutions of the equations

(a) $\sec \theta = 4 \sin \theta$, (b) $\sin \theta + \sin 3\theta = \sin 2\theta$.

(8 marks)

---

7 The table below shows approximate values of a variable y corresponding to certain values of another variable x. It is believed that x and y are related by an equation of the form $y = ax^n$, where a and n are constants. By plotting a suitable graph, show that this is so and estimate the values of a and n to one decimal place.

x	1	2	4	8	15
y	3·8	12·5	41·4	135	398

(8 marks)

---

8 Given that
$$f(x) \equiv x^3 + kx^2 + 9x + 13$$

and $(x + 1)$ is a factor of $f(x)$, find the value of the constant k.

Hence show that the equation $f(x) = 0$ has one real and two complex roots.

Exhibit these roots on an Argand diagram.

(8 marks)

---

9 Show that an equation of the tangent at the point $P(ct, c/t)$, where $t > 0$, to the curve $xy = c^2$, where $c > 0$, is

$$x + t^2 y - 2ct = 0.$$

This tangent meets the x-axis and the y-axis at A and B respectively, and the line PO meets the curve again at Q. The straight line BQ meets the x-axis at L. Prove that the area of triangle AOB is $2c^2$.

Find the area of the triangle QOL.

(9 marks)

---

10 Show, graphically or otherwise, that the equation

$$e^x = 2 - x$$

has only one real root. Using 0·4 as a first approximation to this root, use the Newton–Raphson iterative process to calculate this root to three decimal places.

(9 marks)

---

11 Find the sum of the first n terms of the series

$$x + 3x^2 + 7x^3 + \ldots + (2^r - 1)x^r + \ldots$$

for all values of x, including $x = 1$ and $x = \frac{1}{2}$. State the set of values of x for which the series converges and find its sum to infinity in this case.

(12 marks)

---

12 Given that

$$(1 + x^2)\frac{dy}{dx} = x(1 - y^2)$$

and $y = 0$ when $x = 1$, express y in terms of x.

Sketch the graph of y against x.

(13 marks)

---

13   Evaluate the integrals

(a) $\displaystyle\int_{-1}^{1} \frac{x}{(x+2)^3}\,dx,$ (b) $\displaystyle\int_{0}^{2} \sqrt{(4-x^2)}\,dx,$

(c) $\displaystyle\int_{1}^{e} \ln x\,dx.$

(13 marks)

---

14   Sketch the curve $y = x\,e^{-x}$, showing its asymptote, its maximum point and its point of inflexion.

   Find the area of the finite region R enclosed by the curve, the x-axis and the line $x = 1$. Find also the volume of the solid generated when the region R is rotated through 2π about Ox.

(14 marks)

6.5   Solutions and mark scheme for Paper 6

1          $\overrightarrow{AB} = -\mathbf{i} + 2\mathbf{j} + 4\mathbf{k},\qquad \overrightarrow{AC} = -2\mathbf{i} - 2\mathbf{j} + 2\mathbf{k}.$ **(B1)**

$\overrightarrow{AB} \cdot \overrightarrow{AC} = (2 - 4 + 8) = |\overrightarrow{AB}||\overrightarrow{AC}| \cos BAC.$ **(M1)**

$|\overrightarrow{AB}| = \sqrt{[(-1)^2 + 2^2 + 4^2]} = \sqrt{21},$

$|\overrightarrow{AC}| = \sqrt{[(-2)^2 + (-2)^2 + 2^2]} = \sqrt{12}$ **(B1)**

$\Rightarrow\quad \cos BAC = \dfrac{6}{\sqrt{21}\,\sqrt{12}} = \dfrac{1}{\sqrt{7}}.$ **(A1 cao)**

$\text{Area} = \tfrac{1}{2}|\overrightarrow{AB}||\overrightarrow{AC}| \sin BAC$

$= \tfrac{1}{2}\sqrt{21}\,\sqrt{12}\,\sqrt{\tfrac{6}{7}} = 3\sqrt{6}.$ **(M1 A1)**

---

2   The equations of the circles are

$$(x - 3)^2 + (y + 4)^2 = 16,$$
$$(x + 2)^2 + (y - 8)^2 = 81.$$

(M1)

The centres A, B are $(3, -4)$ and $(-2, 8)$ respectively   **(A1)**

$\Rightarrow\quad AB = \sqrt{[(3 + 2)^2 + (-4 - 8)^2]} = \sqrt{(25 + 144)} = \sqrt{169} = 13.$

But the radii of the circles are 4 and 9 and so the sum of their radii equals the distance between their centres,

$\Rightarrow$ the circles touch externally.   **(A1 cao)**

Let $P \equiv (x, y)$ be a point on the circle on AB as diameter. Then the gradients of PA, PB are $(y + 4)/(x - 3)$, $(y - 8)/(x + 2)$ respectively. **(B1)**

Since $PA \perp PB$, the product of these gradients is -1

$$\Rightarrow \quad \frac{(y + 4)}{(x - 3)} \frac{(y - 8)}{(x + 2)} = -1$$

$$\Rightarrow \quad (x + 2)(x - 3) + (y + 4)(y - 8) = 0 \qquad \textbf{(M1 A1)}$$

is an equation of the circle.

---

3 (a) When R is the middle letter we have to permute 6 letters, of which 2 are D, 2 are S, and 2 are distinct (A and E)

$$\Rightarrow \quad \text{number of ways} = {}^6P_6/({}^2P_2 \times {}^2P_2) = \frac{6!}{4} = 180.$$

(M1 A1 cao)

(b) The total number of permutations is

$$^7P_7/({}^2P_2 \times {}^2P_2) = 1260.$$

The letter R comes first in ${}^6P_6/({}^2P_2 \times {}^2P_2)$ ways = 180 ways. Similarly, R comes last in 180 ways. Therefore

the total number of ways with R neither first nor last is

$$(1260 - 2 \times 180) \text{ ways} = 900 \text{ ways.} \qquad \textbf{(General M2)}$$

(A2, 1 or 0)

---

4 (a) $$\dfrac{(1 + 2x^3)1 - x \cdot 6x^2}{(1 + 2x^3)^2} \qquad \textbf{(M1 A1)} \qquad \Rightarrow \quad -\tfrac{1}{3}. \quad \textbf{(A1 cao)}$$

(b) $$\dfrac{1}{3[1 + (x/3)^2]} \qquad \textbf{(M1 A1)} \qquad \Rightarrow \quad \tfrac{3}{10}. \quad \textbf{(A1 cao)}$$

---

5 $$\dfrac{x(2 + x)}{2x + 1} > 0 \quad \Rightarrow \quad -2 < x < -\tfrac{1}{2} \quad \textbf{(1)} \qquad \text{or} \qquad 0 < x. \qquad \textbf{(1) (1)}$$

$$\dfrac{x(2 + x)}{2x + 1} < 1 \quad \Rightarrow \quad \dfrac{x(2 + x)}{2x + 1} - 1 < 0 \qquad \textbf{(M1)}$$

$$\Rightarrow \quad \dfrac{2x + x^2 - 2x - 1}{2x + 1} < 0 \quad \Rightarrow \quad \dfrac{(x - 1)(x + 1)}{(2x + 1)} < 0 \qquad \textbf{(A1)}$$

$$\Rightarrow \quad x < -1 \quad \textbf{(1)} \qquad \text{or} \qquad -\tfrac{1}{2} < x < 1. \quad \textbf{(1)} \qquad \textbf{(2)}$$

The ranges common to (1) and (2) are $-2 < x < -1$ and $0 < x < 1$. The required set is $\{x : -2 < x < -1\} \cup \{x : 0 < x < 1\}$. **(1 FT)**

---

6 (a)
$$4 \sin \theta \cos \theta = 1$$
$$\Rightarrow \quad 2 \sin 2\theta = 1 \qquad \textbf{(M1)}$$
$$\Rightarrow \quad \sin 2\theta = \tfrac{1}{2}$$
$$\Rightarrow \quad 2\theta = n\pi + (-1)^n \pi/6 \qquad \textbf{(A1 A1)}$$
$$\Rightarrow \quad \theta = n\pi/2 + (-1)^n \pi/12.$$

(b)
$$\sin \theta + \sin 3\theta \equiv 2 \sin 2\theta \cos \theta. \qquad \textbf{(M1)}$$
Therefore
$$2 \sin 2\theta \cos \theta - \sin 2\theta = 0$$
$$\Rightarrow \quad (2 \cos \theta - 1) \sin 2\theta = 0 \qquad \textbf{(M1)}$$
$$\Rightarrow \quad 2 \cos \theta - 1 = 0 \quad \text{or} \quad \sin 2\theta = 0 \qquad \textbf{(A1)}$$
$$\Rightarrow \quad \theta = 2n\pi \pm \pi/3 \quad \text{or} \quad 2\theta = n\pi \ (\Rightarrow \theta = n\pi/2). \textbf{(A1 A1)}$$

7
$$\lg y = \lg a + n \lg x. \qquad \textbf{(M2)}$$

x	$\lg x$	y	$\lg y$
1	0	3·8	0·580
2	0·301	12·5	1·097
4	0·602	41·4	1·617
8	0·903	135	2·130
15	1·176	398	2·600

(Table A1)

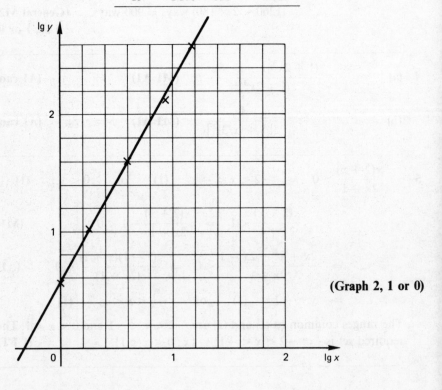

(Graph 2, 1 or 0)

Graph is a straight-line $\Rightarrow$ linear relation between $\lg y$ and $\lg x$ $\Rightarrow$ law is satisfied. **(B1)**

The gradient of the graph of $\lg y$ against $\lg x$ is approximately 1.7, which is the value of n. The graph cuts the $\lg y$ axis where **(A1)**

$$\lg y = 0.59 \quad \Rightarrow \quad \lg a \approx 0.59 \quad \Rightarrow \quad a \approx 3.9. \qquad \textbf{(A1)}$$

---

8 $\qquad$ $f(-1) = 0$ $\quad$ **(M1)** $\qquad \Rightarrow \quad -1 + k - 9 + 13 = 0$ $\quad$ **(A1)**

$$\Rightarrow \quad k = -3. \qquad \textbf{(A1 cao)}$$

By long division,

$$f(x) \equiv (x + 1)(x^2 - 4x + 13). \qquad \textbf{(A1)}$$

The roots are -1 and $\dfrac{4 \pm \sqrt{[(-4)^2 - 4 \times 1 \times 13]}}{2}$,

i.e. $\qquad\qquad -1 \quad$ **(A1)** $\quad$ and $\quad 2 \pm 3i.$ $\quad$ **(A1)**

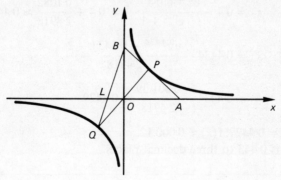

$\qquad\qquad\qquad\qquad\qquad\qquad\qquad\qquad$ **(A2, 1 or 0 FT)**

---

9 $\qquad\qquad\qquad \dfrac{dy}{dx} = \dfrac{-c/t^2}{c} = -\dfrac{1}{t^2}.$ $\qquad\qquad$ **(M1)**

Tangent is $(y - c/t) = -\dfrac{1}{t^2}(x - ct)$

$$\Rightarrow \quad t^2 y - ct + x - ct = 0$$

$$\Rightarrow \quad x + t^2 y - 2ct = 0. \qquad \textbf{(M1 A1 cao)}$$

A is $(2ct, 0)$, B is $(0, 2c/t)$. **(Both A1 cao)**

$$\Delta = \tfrac{1}{2} . \ OA . \ OB = \tfrac{1}{2} . \ 2ct . \ 2c/t = 2c^2.$$ **(A1 cao)**

Q is $(-ct, -c/t)$. **(1)**

BQ has equation $\dfrac{y - 2c/t}{2c/t - (-c/t)} = \dfrac{x}{-(-ct)}$

$$\Rightarrow \quad y - 2c/t = 3x/t^2$$

$$\Rightarrow \quad L \text{ is } (-2ct/3, 0).$$ **(M1 A1)**

$\triangle QOL$ has area $\tfrac{1}{2}|OL| \times |y_Q| = \tfrac{1}{2} . \ 2ct/3 . \ c/t = c^2/3$. **(A1 cao)**

10

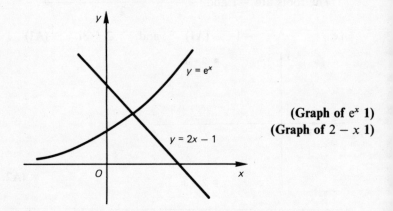

(Graph of e^x 1)
(Graph of $2 - x$ 1)

Clearly the graphs of e^x and $2 - x$ only intersect in one point and so the equation

$$e^x = 2 - x \qquad \textbf{(M2)}$$

has only one real root. Apply the N–R method to $f(x) = 0$, where $f(x) = e^x + x - 2$, with $x_1 = 0{\cdot}4$, $f'(x) = e^x + 1$, **(M1)**

$$x_2 = 0{\cdot}4 - \frac{e^{0\cdot4} + 0{\cdot}4 - 2}{e^{0\cdot4} + 1} \approx 0{\cdot}4 + \frac{0{\cdot}1082}{2{\cdot}4918} \approx 0{\cdot}4434, \qquad \textbf{(A1)}$$

$$x_3 = 0{\cdot}4434 - \frac{e^{0\cdot4434} + 0{\cdot}4434 - 2}{2{\cdot}4918}$$

$$\approx 0{\cdot}4434 - \frac{0{\cdot}001\ 39}{2{\cdot}4918} \approx 0{\cdot}4429. \qquad \textbf{(A1)}$$

When $x = 0{\cdot}4429$, $f(x) \approx 0{\cdot}000\ 12$. **(A1)**
So root is $0{\cdot}443$ to three decimal places. **(A1 cao)**

---

11 Series $= S = \sum_{r=1}^{n} 2^r x^r - \sum_{r=1}^{n} x^r$.   **(M1)**

When $x \neq 1$, $x \neq \frac{1}{2}$,

$$S_n = \frac{2x - 2^{n+1}x^{n+1}}{1 - 2x} - \left(\frac{x - x^{n+1}}{1 - x}\right).$$ **(M1 A1 A1)**

When $x = 1$,

$$S_n = \frac{2 - 2^{n+1}}{1 - 2} - n = 2^{n+1} - (n + 2).$$ **(M1 A1)**

When $x = \frac{1}{2}$,

$$S_n = n - \left(\frac{\dfrac{1}{2} - \dfrac{1}{2^{n+1}}}{1 - \dfrac{1}{2}}\right) = n - 1 + (\tfrac{1}{2})^n.$$ **(M1 A1)**

Series converges when $|2x| < 1 \Rightarrow |x| < \frac{1}{2}$.   **(M1 A1)**

$$S_\infty = \frac{2x}{1 - 2x} - \frac{x}{1 - x}.$$ **(A1 A1)**

---

12 Separating the variables,

$$\frac{dy}{1 - y^2} = \frac{x\,dx}{1 + x^2}$$ **(M1)**

$$\Rightarrow \int \frac{1}{1 - y^2}\,dy = \int \frac{x}{1 + x^2}\,dx + C.$$ **(Constant B1)**

$$\frac{1}{1 - y^2} \equiv \frac{1}{2}\left(\frac{1}{1 - y} + \frac{1}{1 + y}\right)$$ **(M1)**

$$\Rightarrow \int \frac{1}{1 - y^2}\,dy = \frac{1}{2}\ln\left(\frac{1 + y}{1 - y}\right).$$ **(A1)**

Therefore

$$\frac{1}{2}\ln\left(\frac{1 + y}{1 - y}\right) = \frac{1}{2}\ln(1 + x^2) \quad \textbf{(A1)} \qquad + C.$$

When $x = 1$, $y = 0$

$$\Rightarrow \quad C = -\frac{1}{2}\ln 2.$$ **(A1)**

Therefore

$$\frac{1+y}{1-y} = \frac{(1+x^2)}{2} \quad \Rightarrow \quad 2 + 2y = 1 + x^2 - y - x^2 y \qquad \textbf{(M1)}$$

$$\Rightarrow \qquad y = \frac{x^2 - 1}{x^2 + 3}. \qquad \textbf{(A1 cao)}$$

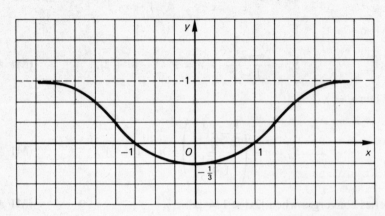

Even function 1)
(Asymptote $y = 1$ 1)
(Intersection with axes 1)
($-\frac{1}{3} < y < 1$ 1)
(General shape 1)

---

13 (a) $\qquad \displaystyle\int_{-1}^{1} \left(\frac{1}{(x+2)^2} - \frac{2}{(x+2)^3} \right) dx \qquad$ **(M1 A1)**

$$= \left[-\frac{1}{(x+2)} + \frac{2}{2(x+2)^2} \right]_{-1}^{1} \qquad \textbf{(Both FT A1)}$$

$$= -\tfrac{1}{3} + \tfrac{1}{9} + 1 - 1 = -\tfrac{2}{9}. \qquad \textbf{(A1 cao)}$$

(b) $\qquad x = 2\sin\theta \quad \Rightarrow \quad dx = 2\cos\theta \, d\theta, \qquad$ **(A1)**

$$\sqrt{(4 - x^2)} = \sqrt{(4 - 4\sin^2\theta)} = 2\cos\theta. \qquad \textbf{(A1)}$$

$$x = 0 \quad \Rightarrow \quad \theta = 0, \qquad x = 2 \quad \Rightarrow \quad \theta = \pi/2. \qquad \textbf{(A1)}$$

$$I = \int_{0}^{\pi/2} 4\cos^2\theta \, d\theta = 2 \int_{0}^{\pi/2} (1 + \cos 2\theta) \, d\theta \qquad \textbf{(M1)}$$

$$= 2\left[\theta + \tfrac{1}{2}\sin 2\theta \right]_{0}^{\pi/2} = \pi. \qquad \textbf{(A1 cao)}$$

(c)
$$\int \ln x \, dx = \int \ln x \, \frac{d(x)}{dx} \, dx \qquad \text{(M1)}$$

$$= x \ln x - \int x \, \frac{d}{dx} (\ln x) \, dx \qquad \text{(A1)}$$

$$= x \ln x - \int x \, \frac{1}{x} \, dx$$

$$= x \ln x - x. \qquad \text{(A1 cao)}$$

$$I = \left[x \ln x - x \right]_1^e = (e \ln e - e - 1 \ln 1 + 1) = 1. \qquad \text{(A1 cao)}$$

---

14 As $x \to \infty$, $y \to 0+$ $\Rightarrow$ asymptote $y = 0$ (with curve above Ox). **(1)**
As $x \to -\infty$, $y \to -\infty$. **(1)**
Also, curve passes through O and $y > 0$ for $x > 0$, $y < 0$ for $x < 0$. **(1)**

$$\frac{dy}{dx} = (1 - x) e^{-x}, \qquad \frac{d^2 y}{dx^2} = (x - 2) e^{-x}$$

$\Rightarrow$ max. at $(1, e^{-1})$, $\dfrac{dy}{dx}$ changes from $(+)$ to $(-)$, **(A1 cao)**

inflexion at $(2, 2 e^{-2})$, $\dfrac{d^2 y}{dx^2}$ changes from $(-)$ to $(+)$. **(A1 cao)**

[Above marks dependent on graph.]

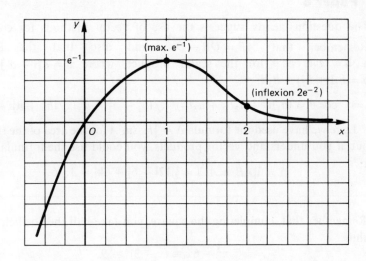

$$\text{Area} = \int_0^1 x \, e^{-x} \, dx \qquad \text{(M1)}$$

$$= \left[-x\,e^{-x} \right]_0^1 + \int_0^1 e^{-x}\,dx \qquad \textbf{(M1)}$$

$$= \left[-(x+1)\,e^{-x} \right]_0^1 = 1 - 2\,e^{-1}. \qquad \textbf{(A1 cao)}$$

$$\text{Volume} = \int_0^1 \pi y^2\,dx \qquad \textbf{(M1)}$$

$$= \pi \int_0^1 x^2\,e^{-2x}\,dx.$$

$$\int x^2\,e^{-2x}\,dx = \int x^2\,\frac{d}{dx}\left(\frac{-e^{-2x}}{2}\right) dx \qquad \textbf{(M1)}$$

$$= -\tfrac{1}{2}x^2\,e^{-2x} + \int x\,e^{-2x}\,dx, \qquad \textbf{(A1)}$$

$$\int x\,e^{-2x}\,dx = -\tfrac{1}{2}x\,e^{-2x} - \tfrac{1}{4}\,e^{-2x}, \qquad \textbf{(M1)}$$

$$\Rightarrow \quad \text{Volume} = \pi \left[-(\tfrac{1}{2}x^2 + \tfrac{1}{2}x + \tfrac{1}{4})\,e^{-2x} \right]_0^1 \qquad \textbf{(A1 cao)}$$

$$= \pi(1 - 5\,e^{-2})/4. \qquad \textbf{(A1 cao)}$$

6.6 Explanation of solutions and mark scheme for Paper 6

1 The question clearly suggests the use of vector methods for cos BAC. Remember that $\overrightarrow{AB} = \overrightarrow{OB} - \overrightarrow{OA}$, and also that the formula $\mathbf{a} \cdot \mathbf{b} = |\mathbf{a}||\mathbf{b}| \cos \theta$ for the scalar product, where $a = a_1\mathbf{i} + a_2\mathbf{j} + a_3\mathbf{k}$, $\mathbf{b} = b_1\mathbf{i} + b_2\mathbf{j} + b_3\mathbf{k}$,

$$\Rightarrow \quad \cos \theta = (a_1 b_1 + a_2 b_2 + a_3 b_3)/[\sqrt{(a_1^2 + a_2^2 + a_3^2)}\,\sqrt{(b_1^2 + b_2^2 + b_3^2)}].$$

Here we have used the formula $\Delta = \tfrac{1}{2}bc \sin A$ for the area of the triangle, but, if you understand vector products, you could use the formula

$$\Delta = \tfrac{1}{2}|\overrightarrow{AB} \times \overrightarrow{AC}| = \tfrac{1}{2}|12\mathbf{i} - 6\mathbf{j} + 6\mathbf{k}| = 3\sqrt{6}.$$

---

2 Remember that, completing the square, you take half the coefficient of x, thus:

$$x^2 - 6x \equiv (x - 3)^2 - 9.$$

Also that the equation of the first circle, when expressed in the form

$$(x - 3)^2 + (y + 4)^2 = 9,$$

means that the square of the distance of the point (x, y) on the circle from the point $(3, -4)$, the centre, is 9 units.

We prove that the circles touch externally by using a basic property of

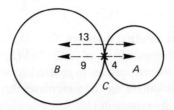

circles, as illustrated by the diagram. Thus, if C is the point on AB distance 4 units from A and 9 units from B, then C must lie on both the circles.

Again, for the second part, we use the basic property of a circle that 'the angle in a semicircle is a right angle' together with the result that, when two lines are perpendicular, the product of their gradients is -1.

Always look for applications of *geometrical properties* of this kind. You could find the equations of the circle by finding its centre, $[\frac{1}{2}(3 - 2), \frac{1}{2}(-4 + 8)]$, and radius, $13/2$, and expressing the equation as

$$(x - \tfrac{1}{2})^2 + (y - 2)^2 = (13/2)^2,$$

but this is less elegant mathematically. However, such a method correctly followed through would gain full marks.

---

3   Remember that a permutation with the two letters D looks the same when these letters are interchanged. Similarly for the two letters S. Also, that n different objects can be arranged in $n!$ ways.

In (b) we derive the answer by finding the total number of permutations of all the letters and subtracting the number of forbidden permutations. (This earns the general **M2**).

---

4   (a) The **(MI)** here is for differentiating the quotient, the **(A1)** for correct working.

Another method would be taking ln and then differentiating thus:

$$\ln y = \ln x - \ln (1 + 2x^3)$$

$$\Rightarrow \quad \frac{1}{y}\frac{dy}{dx} = \frac{1}{x} - \frac{6x^2}{1 + 2x^3} \qquad \textbf{(M1 A1)}$$

$$\Rightarrow \quad -\tfrac{1}{3}. \qquad \textbf{(A1 cao)}$$

Showing working is essential here since, if you quote a wrong answer (with no working), no credit can be awarded. Notice all final answers here are **(cao)**.

(b) The **(M1)** is for differentiating a function of a function *and* use of $\dfrac{d}{dt}(\tan^{-1} t) = 1/(1 + t^2)$, the **(A1)** for correct use in this case.

---

5   The first part of the solution does not show the zeros of the factors in the numerator and denominator of $x(2 + x)/(2x + 1)$, but you should remember to order them from least to greatest (as $-2, -\frac{1}{2}, 0$) and use the fact that the expression changes sign as x increases through each of these values thereby obtaining the ranges in (1).

To consider the right-hand side of the inequation, express it in the form $g(x) < 0$ and repeat the previous process to obtain the ranges in (2).

The final result is obtained by selecting the ranges of x common to (1) and (2), using a number line if necessary, and then expressing the solution in set notation.

You may find the use of number lines very helpful here.

Check your solution with a few trial points, say when $x = -3, -1\frac{1}{2}, \frac{1}{2}, 2$. No marks would be awarded for this check, but your confidence is increased when (if) the check works!

---

6   (a) The solution depends essentially on recognizing that $\sin \theta \cos \theta \equiv \frac{1}{2} \sin 2\theta$. **(M1)**

Then **(A1)** is given for the $n\pi$ and another **(A1)** for the $(-1)^n \pi/6$.

Do check that your solution, at this stage, is correct.

(b) Once more, use of a trigonometrical identity is essential. The $\sin 2\theta$ on the right-hand side of the given equation provides the clue.

Notice the factors $(2 \cos \theta - 1)$ and $\sin 2\theta$ each of which equated to zero gives a set of solutions.

Notice the **(A1)** for *each* of the two parts of the final answer.

---

7   As is standard technique in questions of this type, we take logs and plot lg y against lg x, giving a straight line. A calculator must be used to form the table of values (here given to three decimal places, although two decimal places would be sufficient).

Do not give values to six or seven decimal places. You have to plot a graph!

Note the large scale used so as to obtain values of n and lg a as accurately as possible.

Note also that lg x means $\log_{10} x$.

You could have used ln x etc. if your calculator has the appropriate functions. Remember, accurate use of your calculator is essential here.

---

8 We use the factor theorem to find k.

The quadratic factor can be found by long division or equating coefficients in the identity

$$x^3 - 3x^2 + 9x + 13 \equiv (x + 1)(x^2 + ax + b)$$

(or, for example, putting $x = 0$ and 1 in succession in this identity). Note that the coefficient of x^2 in the quadratic must be 1.

We solve the quadratic equation by the formula and simplify the results. A check on the solution is obtained by multiplying out $(x - 2 - 3i)$ $(x - 2 + 3i)$.

Do not forget the root -1. (Most candidates do!)

In the Argand diagram, one mark (up to a maximum of two) would be lost for each incorrectly (or omitted) plotted root. Note the follow through marks for wrongly calculated roots.

---

9 Remember that, when $x = x(t)$, $y = y(t)$, then $\dfrac{dy}{dx} = \dfrac{dy}{dt} \bigg/ \dfrac{dx}{dt}$. When the question is clearly in parametric form, use the above relation to obtain the gradient of the tangent in terms of t. *Do not* eliminate t and work in cartesian coordinates.

Although a diagram is not essential, it is helpful for you to 'see' where the various named points lie.

The symmetry of the figure enables you to see at once that $OP = OQ$, and that Q is the point with parameter $(-t)$.

Since area is essentially positive, the final answer, for $\triangle QOL$, must be positive.

---

10 Remember that solutions of the equation $h(x) = g(x)$ are given by the x-coordinates of the intersections of the curves $y = h(x)$ and $y = g(x)$, so we draw the curve $y = e^x$, and the line $y = 2 - x$. It is justifiable to say, 'Clearly (in this case), the curves intersect in only one point.'

An alternative proof that the equation has only one root is to say that, 'as x increases from $-\infty$ to ∞, then e^x increases from 0 to ∞ and $2 - x$ decreases from ∞ to $-\infty$ and so the functions can only be equal for one value of x.

Or, if $f(x) = e^x + x - 2$, then $f'(x) > 0$ for all x, so that $f(x)$ increases (from $-\infty$ to ∞) as x increases from $-\infty$ to ∞, and so can only pass through any value (including zero) once.

Notice that we apply the N–R method to the equation $f(x) = 0$, and that we show the values of $f(x_1)$, $f'(x_1)$ in the first iteration

$$x_2 = x_1 - \frac{f(x_1)}{f'(x_1)}.$$

Also that, since $f(x_2)$ is very small, we do not bother to recalculate the denominator $f'(x_2)$ in the second iteration (see p. 112).

Finally, we check our answer by calculating $f(x_3)$, which we find to be of order 10^{-4}.

A calculator is essential in problems of this type. You will find it much faster and more accurate than using tables.

---

11 Notice the clue to the nature of the series (the difference of two geometrical progressions) is given by the displayed rth term.

The sum of the GP

$$ar + ar^2 + \ldots + ar^n = ar(1 + r + r^2 + \ldots + r^{n-1}) \qquad (1)$$

is $ar(1 - r^n)/(1 - r)$, provided $r \neq 1$.

When $r = 1$, the sum is clearly na. This fact is used for the special cases when $x = 1$ and $x = \frac{1}{2}$. Remember to put $x = 1$ (or $\frac{1}{2}$) in both series! As $n \to \infty$ the GP (1) converges to $a/(1 - r)$, provided $|r| < 1$.

---

12 Notice the **(B1)** mark for introducing the constant C. Also, when we have a relation of the form

$$\tfrac{1}{2} \ln f(y) = \tfrac{1}{2} \ln g(x) + \tfrac{1}{2} \ln a,$$

then we can remove the logs and obtain

$$f(y) = ag(x).$$

The graph has been drawn without explanation but, as you see, earns full marks because it correctly displays all the salient features required.

---

13 (a) You cannot integrate the expression as stated, but, expressing x in the form

$$x \equiv \lambda(x + 2) + \mu \quad (\Rightarrow \quad \lambda = 1, \mu = -2),$$

leads to terms which can be integrated on sight.

Remember that integration consists essentially of reducing the integrand to terms which can be integrated at sight.

(b) The surd must be removed and the trigonometric substitution (a well-known one) does this.

Remember that use of simple trigonometric identities, in this case

$$\cos^2 \theta \equiv \tfrac{1}{2}(1 + \cos 2\theta),$$

helps us evaluate integrals with *even* powers of $\sin \theta$ and/or $\cos \theta$.

You could, of course, express the final answer (indefinite integral) in terms of x and substitute limits, but changing the limit is preferable.

(c) Integration by parts is the usual method when a transcendental function occurs in the integrand (perhaps combined with powers of x) and you cannot integrate on sight.

Remember to integrate by parts in the correct direction.

In all three cases here, check, by differentiation, that your indefinite integral is correct.

---

14 Remember, the word 'sketch' means that accuracy of scales, detailed plotting of points is not required (see p. 10), but that just the general shape of the curve with an indication of its significant features is required.

Note that, although the marks for the various deductions regarding the properties of the curve have been given alongside those deductions, it is nevertheless essential that a graph be drawn. The question does say 'Sketch the curve...', and if no graph is drawn, but all the rest is correct, most (if not all) the marks for the sketch will be lost.

Remember that an inflexion on a curve is a point at which the gradient has a maximum or a minimum value.

Throughout this question, values for the coordinates of the maximum point and the point of inflexion, the area and the volume have been left in terms of e. If a numerical value is required, the phrase 'to (so many) significant figures (or decimal places)' will be inserted in the question by the examiners.

The **(M1)** marks for the area and volume integrals require both integrand and limits, although $\int_0^1 \pi y^2 \, dx$ is sufficient for the volume. Remember that $(e^{-x})^2$ is e^{-2x}, *not* e^{-x^2} or e^{x^2}. To evaluate $\int_0^1 x^2 e^{-2x} \, dx$ you have to integrate by parts twice.

6.7 Paper 7 (complete paper with choice)

Time allowed: $2\frac{1}{2}$ hours

Answer ALL questions in Section I and FOUR questions in Section II.

In calculations you are advised to show all the steps in your working, giving your answer at each stage.

SECTION I

(48 marks, 6 marks per question)

1 The functions f and g with domains $\{x : x \in \mathbb{R}, x \neq 0\}$ are defined as follows:

$$f : x \mapsto \frac{1}{x}, \qquad g : x \mapsto x - 1 - \frac{6}{x}.$$

Find constants a and b so that the composite function $h = fg$ is defined on

$$\{x : x \in \mathbb{R}, x \neq 0, x \neq a, x \neq b\}.$$

Express $h(x)$ in partial fractions.

---

2 Find the set of values of x for which $|x - 2| > 5|x - 3|$.

---

3 How many different arrangements can be made with the letters of the word FORTRAN?

In how many of these arrangements are

(a) the vowels together,

(b) the vowels separated by three other letters?

---

4 Find the sum of the first n terms of the series whose rth term is $2^r + 2r - 2$.

---

5 Find

$$(a) \quad \frac{d}{dx}\left(x - \frac{1}{x}\right)^{-2}, \qquad (b) \quad \int \frac{1 + x}{1 + 2x}\,dx.$$

---

6 Solve the equation

$$\sin\left(\frac{\pi}{6} + x\right) - \cos\left(\frac{\pi}{4} + x\right) = 0, \quad \text{for } 0 \leqslant x \leqslant 2\pi.$$

---

7 The table below gives corresponding values of the variables x and y:

x	1	2	3	4	5
y	0·69	0·76	0·83	0·91	1·00

Verify graphically that these values of x and y satisfy approximately a relationship of the form $ay^2 = b^x$, where a and b are constants. From your graph obtain approximate values for a and b.

8 Sketch the curve $y = \sin^2 x$ for $0 \leqslant x \leqslant \pi$.
Show that the area of the region contained between this curve and the line $y = \dfrac{1}{4}$ is $\dfrac{\pi}{6} + \dfrac{\sqrt{3}}{4}$.

SECTION II

(52 marks, 13 marks per question)

9 Find the turning points of the curve whose equation is $y = x/(1 + x^2)$ and show that the origin is a point of inflexion. Sketch the curve showing the behaviour of y when $|x|$ is large.

Calculate the area of the region defined by the inequalities $0 \leqslant x \leqslant a$, $0 \leqslant y \leqslant \dfrac{x}{1 + x^2}$. Show that this area has a value of 2 when $a^2 = e^4 - 1$.

10 The points A, B and C have position vectors $\mathbf{i} + 4\mathbf{j} - 3\mathbf{k}$, $2\mathbf{i} + \mathbf{j} + 2\mathbf{k}$ and $\mathbf{i} + 2\mathbf{j} - \mathbf{k}$ respectively. Find

(a) the vector $\overrightarrow{AB}$,

(b) a vector equation of the straight line AB,

(c) a vector which is perpendicular to AB and AC,

(d) a vector equation of the plane Π containing the points A, B and C,

(e) the distance of the point $D\,(-3, 2, 3)$ from the plane Π.

11 Given that $f(x) \equiv \dfrac{2x + 3}{(1 - x)(4 + x^2)}$, express $f(x)$ in partial fractions.

(a) Find $\displaystyle\int f(x)\,\mathrm{d}x$.

(b) Given that $|x| < 1$, express $f(x)$ as a series in ascending powers of x up to and including the term in x^4.

12 (a) Tabulate, to four significant figures, the values of $f(x)$, where
$f(x) = \dfrac{1}{\sqrt{(1 - x^2)}}$ for values of x equal to 0, 0·1, 0·2, 0·3, 0·4 and 0·5.

Hence, using the trapezoidal rule and the above values of $f(x)$, determine an approximate value of $I = \displaystyle\int_0^{1/2} \dfrac{1}{\sqrt{(1 - x^2)}}\,dx$.

Using this value of I and direct integration find an approximate value for π.

(b) Solve the differential equation $2\dfrac{dy}{dx} + xy^2 = y^2$, given that $y = 1$ when $x = 1$.

---

13 Express $4\cos\theta - 3\sin\theta$ in the form $R\cos(\theta + \alpha)$. Hence,

(a) solve the equation $4\cos\theta - 3\sin\theta = 3$, for $0° \leqslant \theta \leqslant 360°$,

(b) find the greatest and least values of

(i) $(4\cos\theta - 3\sin\theta + 6)^2$, (ii) $\dfrac{1}{4\cos\theta - 3\sin\theta + 6}$.

---

14 The chord of a circle divides the circle into two segments whose areas are in the ratio 1:5. Show that the angle $2x$ radians subtended by the chord at the centre of the circle is given by the equation $\sin 2x = 2x - \pi/3$.

Using the same axes, sketch the graphs, for $0 \leqslant x \leqslant \pi/2$, of $y = \sin 2x$ and $y = 2x - \pi/3$, showing that a root of the equation $\sin 2x = 2x - \pi/3$ is near $x = 1\cdot0$ radian.

Using $x = 1\cdot0$ radian as a first approximation, apply the Newton–Raphson process once to the equation $\sin 2x - 2x + \pi/3 = 0$, to find a second approximation, giving your answer to two significant figures.

7 Longer tests: applied mathematics

We give a detailed, complete marking scheme for every question in Paper 8 in order to explain further how solutions are marked. In order to make the best use of this scheme, you need to try and work the paper first *yourself*. Then, you will need to spend time to master the solution to every question and how it is marked.

7.1 Paper 8

Time allowed: $2\frac{1}{2}$ hours

Answer SIX questions.

1 Prove, by integration, that the centre of mass of a uniform solid hemisphere, of radius r and density ρ, is at a distance $3r/8$ from its plane base.

The hemisphere is made of the same material as a uniform solid circular cylinder of radius r and height $2r$. The circular bases of the hemisphere and cylinder are joined together so that their centres coincide at O. Find the distance from O of the centre of mass G of this composite body.

This body is freely suspended and hangs in equilibrium from a string attached to a point A on the circular edge of the join. Explain why AG is vertical and calculate, to the nearest degree, the inclination of OG to the vertical.

Find, in terms of ρ, g and r, the magnitude of the couple which must be applied to the composite body so that in equilibrium AO is horizontal.

---

2 A particle P moves in a circle, of centre O and radius a, with uniform angular speed ω. Prove that the acceleration of the particle is of magnitude $\omega^2 a$ and is directed along PO.

Particles A and B, of equal mass m, are connected by a light inelastic string, of length l, threaded through a small hole O in a smooth horizontal table. The particle A is free to move on the table and describes a horizontal circle so that OA rotates with constant angular speed ω. The particle B moves below the table, with the string taut, in a horizontal circle with the same angular speed ω. Show that $OA = \frac{1}{2}l$.

If θ is the angle which OB makes with the vertical, find, in terms of g, l and ω, an expression for $\cos \theta$. Show also that $\omega^2 > 2g/l$.

---

3 (a) A particle of mass 2 kg is acted upon by a force **F** which at time t seconds is given by

$$\mathbf{F} = (8\mathbf{i} + 24t^2\mathbf{j})\, \text{N}.$$

When $t = 0$ the particle is at rest and has position vector $(-\mathbf{i} + \mathbf{j})$ m referred to the origin O. Find the velocity and the position vector of the particle when $t = 2$.

When $t \geqslant 2$ the force acting on the particle becomes $8\mathbf{i}$ N. Find the position vector of the particle when $t = 3$.

(b) A particle P is thrown vertically upwards from a point O with speed $2V$ and at the same instant a second particle Q is thrown vertically downwards from a point A, which is at a distance a vertically above O, with speed V. Find the distance from O of the point where P and Q collide.

---

4 One particle A has velocity vector $(3\mathbf{i} - 2\mathbf{j} + 4\mathbf{k})$ m s^{-1} and another particle B has velocity vector $(-\mathbf{i} - 4\mathbf{j} + 2\mathbf{k})$ m s^{-1}. The velocity of a third particle C relative to A is $(-10\mathbf{i} - 4\mathbf{j} + 4\mathbf{k})$ m s^{-1}. Find the velocity vector of C and the velocity of C relative to B.

At time $t = 0$, particle A is at the origin and B is at the point with position vector $(16\mathbf{i} + 8\mathbf{j} + 8\mathbf{k})$ m. Show that A and B will collide and find the position vector of the point of collision.

Given that the mass of A is twice the mass of B and that A and B coalesce on collision, find the velocity of the composite particle after the collision.

---

5 Three perfectly elastic uniform spheres A, B and C, of equal radii but of masses $2m$, m and $3m$ respectively, lie at rest on a smooth horizontal table with their centres in a straight line. Sphere A is projected directly towards sphere B with speed V and after collision sphere B moves directly towards and collides with sphere C.

Prove that after the second collision the kinetic energy of sphere B is twice that of sphere A.

Prove that sphere C will experience only one collision.

---

6   A uniform straight beam AB, of mass m and length $2a$, rests in equilibrium with the end A on rough horizontal ground and the end B against a smooth vertical wall. The beam is in a vertical plane perpendicular to the wall and is inclined at an angle θ to the horizontal. The end A of the beam is connected to the point O of the wall nearest to A by a horizontal spring of natural length a and modulus $\frac{1}{2}mg$.

(a) Calculate the normal reactions at A and B.
(b) Show that the friction force at A is of magnitude $\frac{1}{2}mg(1 + \cot\theta - 2\cos\theta)$ and acts towards O.
(c) Given that the beam is about to slip when $\tan\theta = \frac{3}{4}$, find the coefficient of friction between the beam and the ground.

---

7   (a) An engine of mass 4×10^5 kg works at the rate of 2000 kW while pulling 70 wagons, each of mass 3×10^4 kg, at a steady speed of 36 km h^{-1} along a straight level track. Given that the resistance experienced by each part of the train is proportional to the mass of that part, find, in N, the tension in the coupling between the engine and the first wagon.
     Find the acceleration of the train when the speed is 18 km h^{-1}, if the engine works at the same rate and the resistance remains unaltered.
(b) A pump raises 25 kg of water every second from a depth of 20 m and discharges it with a speed of 10 m s^{-1}. Calculate, in kW, the rate at which the pump is doing useful work. (Take g as 10 m s^{-1}.)

---

8   (a) A particle P is projected from the point O, with speed V at an angle θ to the horizontal plane which passes through O, and moves freely under gravity. Show that the time of flight is $(2V\sin\theta)/g$, that the range on the plane is $(V^2\sin 2\theta)/g$ and that the greatest height of P above the plane is $(V^2\sin^2\theta)/(2g)$.
     Given that the range is less than V^2/g, show that there are two possible acute angles of projection and that, if these angles are θ_1 and θ_2, then

$$\theta_1 + \theta_2 = \pi/2.$$

(b) A particle starts from O with initial velocity 5 m s^{-1} and travels along Ox. At time t seconds the acceleration of the particle in the direction x increasing is $20/(4 + t)^2$ m s^{-2}. Find, to the nearest m, the distance covered by the particle during the first four seconds of the motion.

---

9 A box contains nine balls of which four are white, three are green and two are red. Three balls are to be drawn, one at a time at random and without replacement, from the box. Calculate the probability that

(a) the balls, in the order drawn, will be coloured white, green and red respectively,

(b) all three balls will be of the same colour,

(c) the third ball drawn will be white,

(d) no white ball will be drawn.

7.2 Solutions and mark scheme for Paper 8

Each question is marked out of seventeen marks. An explanation of the solutions and marking scheme follows in § 7.3.

1

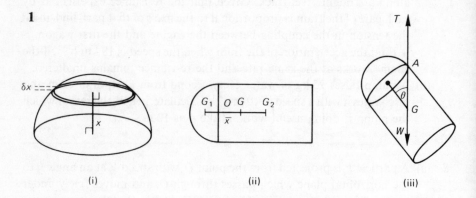

(i) (ii) (iii)

Mass of slice = $\pi(r^2 - x^2)\rho \, \delta x$ (see Fig. i).

Moment of slice about the base = $\pi(r^2 - x^2)\rho \, \delta x \times x$.

$$\bar{x} = \frac{\displaystyle\int_0^r \pi\rho(r^2 x - x^3)\,dx}{\displaystyle\int_0^r \pi\rho(r^2 - x^2)\,dx} = \frac{\left[\dfrac{r^2 x^2}{2} - \dfrac{x^4}{4}\right]_0^r}{\left[r^2 x - \dfrac{x^3}{3}\right]_0^r} = \frac{\dfrac{r^4}{4}}{\dfrac{2r^3}{3}} = \frac{3r}{8}.$$

(M2 A3, 1 or 0)

$$\bar{x} = \frac{2\pi r^3 \rho r - \dfrac{2}{3}\pi r^3 \rho \dfrac{3r}{8}}{2\pi r^3 \rho + \dfrac{2}{3}\pi r^3 \rho} = \frac{21r}{32}$$
(see Fig. ii). **(M2 A2, 1 or 0)**

In equilibrium, the only forces acting on the composite body are its weight, of magnitude W, acting vertically downwards, and the tension, of magnitude T, in the string (see Fig. iii). To balance, these forces must be equal and opposite and so the string along which the tension acts must be vertical. Further, since the tension acts through A and the weight acts through G, A and G must be in the same vertical straight line for the tension and the weight to be equal and opposite.

(3, 2 or 0)

If θ is the required angle,

$$\tan\theta = \frac{OA}{OG} = \frac{32}{21} \;\Rightarrow\; \theta = 57° \text{ (nearest degree).}$$

(M1 A1 cao)

When AO is horizontal, and the magnitude of the required couple is L,

$$\circlearrowleft A, \;\Rightarrow\; L = \rho(2\pi r^3 + \tfrac{2}{3}\pi r^3)gr = 8\pi\rho gr^4/3. \quad \textbf{(M2 A1 cao)}$$

2

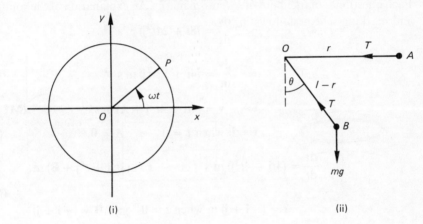

(i)   (ii)

Many proofs are available for this 'bookwork'.

For example: let P be the point $x = a\cos\omega t$, $y = a\sin\omega t$, as shown in Fig. (i).

Then
$$\dot{x} = -a\omega\sin\omega t, \qquad \dot{y} = a\omega\cos\omega t,$$
$$\ddot{x} = -a\omega^2\cos\omega t, \qquad \ddot{y} = -a\omega^2\sin\omega t$$

$\Rightarrow$ acceleration is of magnitude

$$\sqrt{(a^2\omega^4\cos^2\omega t + a^2\omega^4\sin^2\omega t)} = a\omega^2$$

directed along $\overrightarrow{PO}$.

(4, 3 or 0)

In Fig. (ii) let the tension in the string be of magnitude T and the length of the string on the table be r.

Motion of A along AO, $\;\Rightarrow\; T = mr\omega^2.$ (1) **(M1 A1)**

Resolving $\updownarrow$ for B, $\;\Rightarrow\; T\cos\theta = mg.$ (2) **(M1 A1)**

Resolving $\leftrightarrow$ for B, $\Rightarrow$ $T \sin \theta = m(l - r)\omega^2 \sin \theta$

(M1 A1)

$\Rightarrow$ $T = m(l - r)\omega^2.$ (3)

(1) and (3) $\Rightarrow$ $mr\omega^2 = m(l - r)\omega^2$ $\Rightarrow$ $r = \frac{1}{2}l.$

(M1 A1 cao)

Then (1) $\Rightarrow$ $T = \frac{1}{2}ml\omega^2.$ **(A1)**

Then (2) $\Rightarrow$ $\cos \theta = 2g/(l\omega^2).$ **(M1 A1)**

But $\cos \theta \leqslant 1$ and, since B moves in a circle, $\cos \theta \neq 1.$

Therefore $2g/(l\omega^2) < 1$ $\Rightarrow$ $\omega^2 > 2g/l.$ **(M1 A1 cao)**

---

3 (a) For $t \leqslant 2$, applying Newton's Second Law gives

$$2 \frac{d\mathbf{v}}{dt} = (8\mathbf{i} + 24t^2\mathbf{j}) \text{ m s}^{-2}$$

$$\Rightarrow \frac{d\mathbf{v}}{dt} = (4\mathbf{i} + 12t^2\mathbf{j}) \text{ m s}^{-2} \quad \textbf{(M1)}$$

$$\Rightarrow \mathbf{v} = (4t\mathbf{i} + 4t^3\mathbf{j} + \mathbf{A}) \text{ m s}^{-1}. \quad \textbf{(M1 A1)}$$

$$\mathbf{v} = \mathbf{0} \text{ when } t = 0 \Rightarrow \mathbf{A} = \mathbf{0}. \quad \textbf{(A1)}$$

$$\frac{d\mathbf{r}}{dt} = (4t\mathbf{i} + 4t^3\mathbf{j}) \text{ m s}^{-1} \Rightarrow \mathbf{r} = (2t^2\mathbf{i} + t^4\mathbf{j} + \mathbf{B}) \text{ m}.$$

(M1)

$$\mathbf{r} = (-\mathbf{i} + \mathbf{j}) \text{ m when } t = 0 \Rightarrow \mathbf{B} = (-\mathbf{i} + \mathbf{j})$$

$$\Rightarrow \mathbf{r} = \left[(2t^2 - 1)\mathbf{i} + (t^4 + 1)\mathbf{j}\right] \text{ m}. \quad \textbf{(A1)}$$

When $t = 2$,

$$\mathbf{v} = (8\mathbf{i} + 32\mathbf{j}) \text{ m s}^{-1}, \quad \mathbf{r} = (7\mathbf{i} + 17\mathbf{j}) \text{ m}. \quad \textbf{(A1 cao)}$$

For $t \geqslant 2$,

$$\frac{d\mathbf{v}}{dt} = 4\mathbf{i} \text{ m s}^{-2}$$

$$\Rightarrow \mathbf{v} = (4t\mathbf{i} + \mathbf{C}) \text{ m s}^{-1}. \quad \textbf{(M1)}$$

When $t = 2$,

$$\mathbf{v} = (8\mathbf{i} + 32\mathbf{j}) \text{ m s}^{-1} \Rightarrow \mathbf{C} = 32\mathbf{j}$$

$$\Rightarrow \mathbf{v} = (4t\mathbf{i} + 32\mathbf{j}) \text{ m s}^{-1} \quad \textbf{(A1)}$$

$$\Rightarrow \frac{d\mathbf{r}}{dt} = (4t\mathbf{i} + 32\mathbf{j}) \text{ m s}^{-1}$$

$$\Rightarrow \mathbf{r} = (2t^2\mathbf{i} + 32t\mathbf{j} + \mathbf{D}) \text{ m}. \quad \textbf{(M1)}$$

When $t = 2$,

$$\mathbf{r} = (7\mathbf{i} + 17\mathbf{j}) \text{ m} \quad \Rightarrow \quad 7\mathbf{i} + 17\mathbf{j} = 8\mathbf{i} + 64\mathbf{j} + \mathbf{D}$$

$$\Rightarrow \mathbf{D} = (-\mathbf{i} - 47\mathbf{j}) \tag{A1}$$

$$\Rightarrow \mathbf{r} = \left[(2t^2 - 1)\mathbf{i} + (32t - 47\mathbf{j}\right] \text{ m.}$$

When $t = 3$,

$$\mathbf{r} = (17\mathbf{i} + 49\mathbf{j}) \text{ m.} \tag{A1 cao}$$

(b) The relative velocity (of approach) of P and Q has magnitude $3V$, and the relative acceleration is zero. **(M2)**

They meet after time $a/(3V)$. **(A1)**

The height above O of the point where they meet is

$$2V(a/3V) - \tfrac{1}{2}g(a/3V)^2 = 2a/3 - ga^2/(18V^2).$$

(M1 A1 cao)

---

4 $\qquad \mathbf{v}_C - \mathbf{v}_A = (-10\mathbf{i} - 4\mathbf{j} + 4\mathbf{k}) \text{ m s}^{-1}$

$$\Rightarrow \mathbf{v}_C = (3\mathbf{i} - 2\mathbf{j} + 4\mathbf{k}) \text{ m s}^{-1} + (-10\mathbf{i} - 4\mathbf{j} + 4\mathbf{k}) \text{ m s}^{-1}$$

$$= (-7\mathbf{i} - 6\mathbf{j} + 8\mathbf{k}) \text{ m s}^{-1}. \tag{M1 A1}$$

Velocity of C relative to $B = \mathbf{v}_C - \mathbf{v}_B$ **(M1)**

$$= (-7\mathbf{i} - 6\mathbf{j} + 8\mathbf{k}) \text{ m s}^{-1} - (-\mathbf{i} - 4\mathbf{j} + 2\mathbf{k}) \text{ m s}^{-1}$$

$$= (-6\mathbf{i} - 2\mathbf{j} + 6\mathbf{k}) \text{ m s}^{-1}. \tag{M1 A1}$$

At time t seconds,

$$\mathbf{r}_A = (3\mathbf{i} - 2\mathbf{j} + 4\mathbf{k})t \text{ m,} \tag{M1 A1}$$

$$\mathbf{r}_B = (-\mathbf{i} - 4\mathbf{j} + 2\mathbf{k})t \text{ m} + (16\mathbf{i} + 8\mathbf{j} + 8\mathbf{k}) \text{ m}$$

$$= \left[(16 - t)\mathbf{i} + (8 - 4t)\mathbf{j} + (2t + 8)\mathbf{k}\right] \text{ m.} \tag{A1}$$

If A and B collide,

$$\mathbf{r}_A = \mathbf{r}_B \tag{M1}$$

$$\Rightarrow \quad 16 - t = 3t, \ 8 - 4t = -2t, \ 2t + 8 = 4t, \tag{M1 A1}$$

all of which are satisfied when $t = 4$. **(A1 cao)**

They collide when $t = 4$ at the point with position vector

$$(12\mathbf{i} - 8\mathbf{j} + 16\mathbf{k}) \text{ m.} \tag{A1 cao}$$

Using the principle of conservation of linear momentum,

$$3m\mathbf{v} = 2m\mathbf{v}_A + m\mathbf{v}_B, \qquad \text{(M2 A1)}$$

where m kg is the mass of B and $\mathbf{v}$ is the velocity of the composite particle of mass $3m$,

$$\Rightarrow \quad \mathbf{v} = \tfrac{1}{3}(5\mathbf{i} - 8\mathbf{j} + 10\mathbf{k}) \text{ m s}^{-1}. \qquad \text{(A1 cao)}$$

5

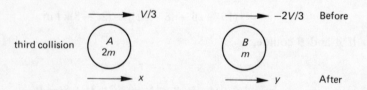

Newton's Law: $\qquad\qquad v - u = -(-V).$ **(M1 A1)**

Linear momentum $\qquad mv + 2mu = 2mV$ **(M1 A1)**

$$\Rightarrow \quad u = V/3, \qquad v = 4V/3. \qquad \text{(A1 cao)}$$

Newton's Law: $\qquad\qquad q - p = -(-4V/3).$

     (M1 A1 A1)

Linear momentum $\qquad 3mq + mp = m(4V/3)$

$$\Rightarrow \quad p = -2V/3, \qquad q = 2V/3. \qquad \text{(A1 cao)}$$

The kinetic energies of spheres A and B are now $\tfrac{1}{2}(2m)(V/3)^2 = mV^2/9$ and $\tfrac{1}{2}m(2V/3)^2 = 2mV^2/9$ respectively and so the kinetic energy of B is now twice that of A.   **(M1 A1 cao)**

Since $p < 0$, sphere B is now moving $\leftarrow$ and there will be another collision between A and B.   **(M1)**

Newton's Law: $\qquad\qquad y - x = -(-2V/3 - V/3).$

     (M1 A1)

Linear momentum $\quad my + 2mx = -2mV/3 + 2mV/3$

$$\Rightarrow \quad x = -V/3, \qquad y = 2V/3. \qquad \text{(A1 cao)}$$

The velocities of the spheres are now

$$\underset{A}{\leftarrow} \ V/3 \qquad \underset{B}{\rightarrow} \ 2V/3 \qquad \underset{C}{\rightarrow} \ 2V/3$$

so there will be no more collisions. Therefore C experiences only one collision.

(A2 cao)

6

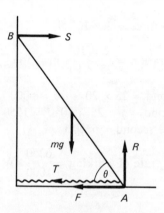

(a) Forces correctly shown on figure. **(3, 2, 1 or 0)**

$$\updownarrow, \ \Rightarrow \ R = mg. \tag{M1}$$

$$\overset{\curvearrowleft}{A}, \ \Rightarrow \ S(2a \sin \theta) = mg(a \cos \theta) \tag{M2}$$

$$\Rightarrow \ S = \tfrac{1}{2}mg \cot \theta. \tag{A1}$$

(b) Hooke's Law

$$\Rightarrow \ T = \tfrac{1}{2}mg(2a \cos \theta - a)/a = \tfrac{1}{2}mg(2 \cos \theta - 1).$$
(M2 A1)

$$\leftrightarrow, \ \Rightarrow \ S = T + F \tag{M1 A1}$$

$$\Rightarrow \ F = \tfrac{1}{2}mg(\cot \theta + 1 - 2 \cos \theta). \tag{A1 cao}$$

(c) When $\tan \theta = \tfrac{3}{4}$, $\cos \theta = 4/5$, $F = 11mg/30$. **(A2)**
In limiting equilibrium,

$$F/R = \mu \ \Rightarrow \ \mu = 11/30. \qquad \textbf{(M1 A1 FT)}$$

---

7 (a) $$\text{Speed} = 36 \times 5/18 \text{ m s}^{-1} = 10 \text{ m s}^{-1}. \qquad \textbf{(A1)}$$

$$\text{Total pull of engine} = \frac{2000 \times 1000}{10} \text{N} = 2 \times 10^5 \text{ N}$$

$$= \text{total resistance.} \qquad \textbf{(M1 A1)}$$

$$\text{Tension in coupling} = \frac{\text{mass of wagons}}{\text{mass of engine and wagons}} \times \text{total pull}$$

$$= \frac{21 \times 10^5}{21 \times 10^5 + 4 \times 10^5} \times 2 \times 10^5 \, \text{N}$$

$$= 1\cdot68 \times 10^5 \, \text{N}. \qquad \textbf{(M2 A1 cao)}$$

When the speed is $18 \, \text{km h}^{-1} = 5 \, \text{m s}^{-1}$, the pull of the engine is $4 \times 10^5 \, \text{N}$

$$\Rightarrow \quad \text{accelerating force} = 2 \times 10^5 \, \text{N} \qquad \textbf{(M1 A1)}$$

$$\Rightarrow \quad \text{acceleration} = \frac{2 \times 10^5}{25 \times 10^5} \, \text{m s}^{-2} = \frac{2}{25} \, \text{m s}^{-2}. \qquad \textbf{(M1 A1 cao)}$$

(b) Gain of PE/second $= 25 \times 20 \times g \, \text{J} = 5000 \, \text{J}.$ **(M1 A1)**
Gain of KE/second $= \frac{1}{2} \times 25 \times 10^2 \, \text{J} = 1250 \, \text{J}.$ **(M1 A1)**
Total work done/second $= 6250 \, \text{J}$ **(A1 FT)**

$$\Rightarrow \quad \text{rate of working} = \frac{6250}{1000} \, \text{kW} = 6\cdot25 \, \text{kW}. \textbf{(M1 A1 cao)}$$

---

8 (a)

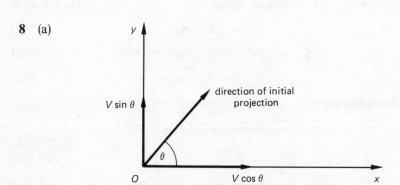

Horizontally, $x = (V \cos \theta)t,$

Vertically, $y = (V \sin \theta)t - \frac{1}{2}gt^2.$ **(A1 both)**

$$y = 0 \text{ when } (V \sin \theta)t - \tfrac{1}{2}gt^2 = 0$$

$$\Rightarrow \quad t = (2V \sin \theta)/g = \textbf{time of flight}. \qquad \textbf{(M1 A1 cao)}$$

Range $= (V \cos \theta) \times (\text{time of flight})$

$$= (V \cos \theta)(2V \sin \theta)/g$$

$$= V^2(2 \sin \theta \cos \theta)/g = (V^2 \sin 2\theta)/g. \qquad \textbf{(M1 A1 cao)}$$

Greatest height = (vertical component of initial velocity)^2/(2g)

$$= (V^2 \sin^2 \theta)/(2g). \qquad \textbf{(M1 A1 cao)}$$

When range is $R < V^2/g$, θ satisfies the equation

$$(V^2 \sin 2\theta)/g = R$$

$$\Rightarrow \quad \sin 2\theta = gR/V^2 < 1.$$

This will have a solution, θ_1 say, between 0 and $\pi/4$. Then

$$\sin\left[2(\pi/2 - \theta_1)\right] = \sin(\pi - 2\theta_1) = \sin 2\theta_1,$$

and so there is a second angle θ_2 $(= \pi/2 - \theta_1)$ of projection. **(M2 A1)**

(b)
$$\frac{dv}{dt} = \frac{20}{(4+t)^2}$$

$$\Rightarrow \quad v = -\frac{20}{4+t} + A. \qquad \textbf{(M1 M1)}$$

$$v = 5 \text{ when } t = 0 \quad \Rightarrow \quad A = 10 \qquad \textbf{(A1)}$$

$$\Rightarrow \quad \frac{dx}{dt} = 10 - \frac{20}{4+t}$$

$$\Rightarrow \quad x = 10t - 20 \ln(4+t) + B, \qquad \textbf{(M1 A1)}$$

$$x = 0 \text{ when } t = 0 \quad \Rightarrow \quad B = 20 \ln 4$$

$$\Rightarrow \quad x = 10t - 20 \ln\left(\frac{4+t}{4}\right). \qquad \textbf{(A1)}$$

When $t = 4$, $x = 40 - 20 \ln 2 \approx 26{\cdot}14$.
The particle covers approximately 26 m in the first four seconds. **(A1)**

---

9 (a) $\qquad P(W, G, R) = \frac{4}{9} \times \frac{3}{8} \times \frac{2}{7} = \frac{1}{21}.$ $\qquad \textbf{(M2 A1)}$

(b) P(all same colour) = $P(W, W, W) + P(G, G, G)$ $\qquad \textbf{(M2)}$

$$= \frac{4}{9} \times \frac{3}{8} \times \frac{2}{7} + \frac{3}{9} \times \frac{2}{8} \times \frac{1}{7} \qquad \textbf{(A1 A1)}$$

$$= \frac{30}{9 \times 8 \times 7}$$

$$= \frac{5}{84}. \qquad \textbf{(A1 cao)}$$

(c) P(third ball is white) = $\frac{4}{9}$. (Since the balls are drawn at random, the probability that the third ball will be white is the same as the probability that a single drawing results in a white ball.) **(M2 A1)**

(d) P(no white ball). A probability tree gives the easiest solution for this, but note that it is not necessary to draw the whole tree.

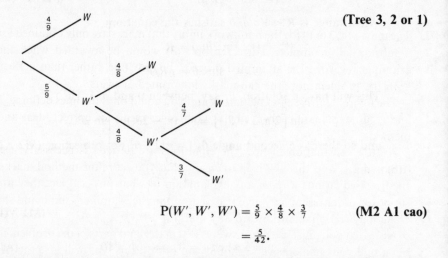

(Tree 3, 2 or 1)

$$P(W', W', W') = \tfrac{5}{9} \times \tfrac{4}{8} \times \tfrac{3}{7}$$

(M2 A1 cao)

$$= \tfrac{5}{42}.$$

7.3 Explanation of solutions and mark scheme for Paper 8

Note the use of abbreviations etc. in the model solutions. You must give sufficient indication of what your symbols stand for but often this can be done in the diagram.

1 The general **(M2)** for deriving the $3r/8$ could be obtained for just writing down the correct expression for $\bar{x}$ in terms of integrals (line 3) but the first two lines (for mass and moment of slice) plus an attempt to sum by integration would also earn the method. By **(A3, 1 or 0)** is meant **(A3)** if fully correct, **(A1)** if there is just one arithmetical slip, **(A0)** if more than one slip. However, no marks can be obtained for integration, no matter how clever, unless you have the correct integrals for this bookwork.

For the position of G, the method **(M2)** is for a valid attempt at a complete solution using a moments equation and notice that this is completely independent of the previous bookwork for which the answer is given should you fail to do it.

Notice carefully the logic used in *proving* that AG is vertical. You must write down the physical reasons and not just agree with the examiner, as many candidates will do without any proof.

A clear, marked diagram will help you in each of the final stages. We have shown the one used for finding θ and left you to sketch one for the determination of L.

Finally, read the question through again to ensure that you have completed all you were asked to do.

---

2   The marks **(4, 3 or 0)** for the bookwork imply that marks are only obtained for a substantially complete, correct proof; **(3)** would be awarded when the examiner regarded the attempted proof to be only fair, rather than perfect.

In the problem note the **cao** marks for printed answers.

The figure is fairly self-explanatory and you could omit the lines defining T and r, but explanation always helps you to get credit for correct ideas and methods.

A little explanation of how the equations are derived is necessary, e.g. 'Resolving ↕,' and this will help to ensure that you score the method marks. Incorrect equations *without* any explanation of what physical fact they are supposed to represent may give rise to doubt in the examiner's mind and she or he may feel unable to award any mark at all.

Note that you must use the answer $r = \frac{1}{2}l$ in the final parts of the problem in obtaining your final answers; should you not do this, you will lose the **(A)** mark in each case.

---

3   (a) Note that t is a number throughout this question. The units have been correctly organized—make sure that you do this also!

Also, integration of $\dfrac{d\mathbf{v}}{dt} = \dots$ involves the introduction of an arbitrary *vector* constant **A**.

The motion changes character when $t > 2$ (the force acting is different) but the new arbitrary constant **C** is determined by equating the two formulae for **v** at $t = 2$. Likewise, the vector constant **D** is determined by equating the two formulae for **r** at $t = 2$.

(b) Since both P and Q have the same acceleration, of magnitude g downwards, their relative velocity of approach remains constant (the value at $t = 0$). Of course, you could consider the distances moved by each in time t and equate their sum to $3a$.

---

4   Again, note the care taken with units in the wording of this question (t once more being just a number).

The solution is wholly dependent on the use of $\mathbf{v}_X - \mathbf{v}_Y$ as the velocity of X relative to Y. Without the use of this, no method (or accuracy) will be earned for the first 5 marks.

To show that A and B collide, we find $\mathbf{r}_A$ and $\mathbf{r}_B$ in terms of t and show that these are equal for a particular value of t. Remember that the equality of two vectors means equality of all three components and the three equations in line 13 are *all* satisfied when $t = 4$.

The final method **(M2)** is earned for recognition of the need for and an attempt to apply the principle of conservation of linear momentum. Make sure you understand the meaning of words like *'coalesce'* and *'composite'* as used here.

---

5 Note the use of a diagram at each collision showing the velocities before and after. This saves explanations of the symbols. Also note the physical explanation for the equations written down, thereby ensuring the method marks. This is a popular topic often examined and you can plan your strategy for such questions. It is advisable to show all velocities in the same direction (a negative sign will imply moving in the opposite direction) and then the two laws can be written thus:

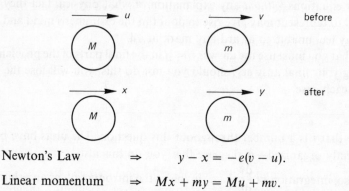

Newton's Law $\Rightarrow$ $y - x = -e(v - u)$.

Linear momentum $\Rightarrow$ $Mx + my = Mu + mv$.

Note the final explanation for only one collision experienced by C. You must give some clear proof and not just agree with the examiner.

---

6 Again, a clear diagram with forces correctly shown, and, throughout, some physical reasons for the equations written down, ensure credit for what you know.

Note that the force of friction is not taken as limiting until part (c). You must not assume $F = \mu R$ unless it is stated that friction is limiting, and parts (a) and (b) do not require this.

Remember T is determined by Hooke's Law and the length of the spring. Also, that just three unknowns (in this case F, R and S) can be determined in a (two-dimensional) statics problem.

Of course, the values of F, R and S can be calculated in other ways, but the equations used in the solution are the simplest.

Don't write down too many equations. If you make this mistake, arithmetical errors will lead you into a mess which you will be unable to sort out in the exam time available.

---

7 (a) Note the need to express the speed of the train in m s^{-1} and that this speed is steady along a straight, level track. This implies that just two forces are acting on the whole train: the pull and the resistance. Further, these forces are equal in magnitude and opposite in direction. Draw a sketch to show this.

At steady speed there are just two forces acting on the wagons taken as a single body, i.e. the tension in the coupling between the first wagon and the engine and the resistance, and you are told how to find the resistance in the question. Support your solution with another sketch.

The final stage of (a) of this question presents a situation which occurs for just an instant. The pull of the engine exceeds the resistance and the excess force being produced causes the train to have an instantaneous acceleration. Draw a sketch to support your solution.

(b) Note that the water is being raised and given speed. By finding the total energy required each second to achieve this, we are able to calculate the rate at which the pump is doing useful work.

In the question as a whole, note the dependence of a correct solution on the units being used in each stage. Train yourself to write in the units each time as shown in this solution.

---

8 Make sure that you are able to write down proofs for the standard formulae asked for in this question quickly and concisely, because they are often requested and they should provide you with some easy marks.

In (b), everything depends on starting correctly with the differential equation where the acceleration is written in the form $\dfrac{dv}{dt}$. The variables in this equation may then be separated and the equation can be integrated to find v in terms of t. Note how the constant resulting from the integration is evaluated from the data given in the question. Train yourself to include a constant of integration always. The remaining stage in the solution then requires us to write v as $\dfrac{dx}{dt}$, separate the variables x and t, integrate, evaluate the constant of integration, and complete as shown. Of course, this solution depends to a very large extent on your ability to integrate correctly and confidently. Make sure that you can do this.

---

9 (a) The method marks (M2) are for correctly getting the probability for each of W, G, R and *multiplying* these probabilities. M1 is for correct probabilities but failure to multiply. For any of the probabilities wrong, M2 for 1 wrong, and M1 for 2 wrong, provided the probabilities are multiplied.

(b) M2 is for P(W, W, W) and P(G, G, G) and adding. Then A1 is given for each of the products correctly given. A1 cao for 5/84.

(c) **M2** is given for seeing how to obtain the probability but also for any other method, e.g. a tree diagram. **A1 cao** for 4/9.

(d) For a probability tree, allow **3** for all correct, **2** for 1 error, **1** for 2 or more errors. For finding the probabilities from the tree diagram, allow **1** for the probabilities correctly found and **1** for the multiplication. **A1 cao** for 5/42.

7.4 Paper 9 (short questions)

Time allowed: $2\frac{1}{2}$ hours

Answer ALL questions.

1   A smooth sphere A, of mass $3m$ and moving with speed V, collides directly with a smooth sphere B, of mass $2m$, which is at rest. After the impact the speed of B is V. Calculate the coefficient of restitution between the spheres.

(5 marks)

2   A uniform solid hemisphere rests in equilibrium with its plane face vertical and its curved surface on a rough inclined plane. Find, correct to the nearest degree, the angle between the plane and the horizontal.

(5 marks)

3   A light spring, obeying Hooke's Law, of natural length l and having one end fixed, has an extension x when the tension in the spring is T. Find, in terms of T, l and x, the modulus of elasticity of the spring. Find also, in terms of T and x, the work done in extending the spring from this position to a length $(l + 2x)$.

(6 marks)

4   A water pump raises water at a rate of 50 kg a second through a height of 20 m. The water emerges as a jet with speed 50 m s^{-1}. Find the kinetic energy and the potential energy given to the water each second and hence find, in kW, the effective power developed by the pump. [Take g as 10 m s^{-2}.] **(6 marks)**

5   A car, moving along a straight level road with constant speed $2u$, passes a point O at the same instant as a second car starts to accelerate with constant acceleration f from rest at O. Show that the distance covered before the cars are level again is $8u^2/f$.

(6 marks)

6   Two forces, of magnitudes 35 N and 40 N, act from a point O in the directions north and south-east respectively. The resultant of these forces is **R**. Calculate

(a) the magnitude of **R**, giving your answer to the nearest N,
(b) the direction of **R**, giving your answer to the nearest degree.

(6 marks)

7   A uniform beam ABC, of weight W and length $3l$, rests horizontally on supports at A and B, where $AB = 2l$. A man of weight $2W$ stands on the beam at a distance x from A. The maximum load that the support at A can bear is $2W$. Show that equilibrium is only possible if $l \leqslant 4x \leqslant 9l$.

(7 marks)

8   Two particles, of masses $2m$ and m, are attached to the ends of a light inextensible string, of length $12a$, which passes over a small smooth light pulley fixed at a height $9a$ above a horizontal plane. The system is released from rest with each particle at a height $3a$ above the plane. Prove that the heavier particle reaches the plane after time $3\sqrt{(2a/g)}$.
    Find the tension in the string during the motion.

(8 marks)

9   A particle P, of mass m, moving on the inner surface of a hollow smooth spherical bowl, of centre O and internal radius $13a$, describes a horizontal circle at a depth $12a$ below O. Find, in terms of m and g, the magnitude of the force exerted by the bowl on P and show that the speed of P is $\frac{5}{2}\sqrt{(ga/3)}$.

(8 marks)

10   The particle P moves from rest at a point A with acceleration proportional to its distance from a fixed point O and always directed towards O. If $OA = a$ and the magnitude of the initial acceleration is k^2a, obtain an expression for the speed of the particle P when $OP = x$.
    Find the time which elapses before P first returns to A and the time taken for P to travel directly from A to the mid-point of OA.

(8 marks)

11 A particle A, moving with constant velocity $(3\mathbf{i} - 2\mathbf{j})$ m s^{-1}, passes through the point with position vector $(4\mathbf{i} + 4\mathbf{j})$ m at the same instant as a particle B passes through the point with position vector $(-4\sqrt{2}\mathbf{i} + p\mathbf{j})$ m. Given that B has a constant velocity $(4\mathbf{i} - \mathbf{j})$ m s^{-1}, find the velocity of B relative to A and the value of p which ensures that A and B collide.

(9 marks)

---

12 A particle is projected with speed V at an angle θ to the horizontal from a point O on a horizontal plane. The particle hits the plane again at the point A. Find the range OA.

If the range OA is equal to the greatest height reached by the particle, show that $\tan \theta = 4$. (Take g as 10 m s^{-2}.)

(9 marks)

---

13 A particle, of mass 2 kg, is free to move on the surface of a smooth horizontal table and is acted upon by a variable force $\mathbf{F}$, where

$$\mathbf{F} = (4\,e^{2t}\mathbf{i} + 16\,e^{-2t}\mathbf{j})\ \text{N}$$

at time t seconds. The particle starts when $t = 0$ with velocity $(2\mathbf{i} + 3\mathbf{j})$ m s^{-1}. Find the speed of the particle when $t = \ln 2$.

(9 marks)

---

14 Bag A contains 4 green balls and 2 black balls and balls are drawn at random, with replacement. Find the probability that the third ball drawn is the second black ball to be drawn.

Two additional bags, labelled B and C, are such that bag B contains 3 green balls and 5 black balls, while bag C contains 2 green balls and 1 black ball. One of the three bags A, B and C is selected at random and a ball is drawn from this bag. Find the probability that

(a) bag B is selected and a green ball is drawn,
(b) a green ball is drawn.

(10 marks)

7.5 Solutions and mark scheme for Paper 9
You should draw your own diagrams for this paper.

1 Momentum conserved: $3mV = 3mV' + 2mV$, where V' is speed of A after the collision, **(M1)**

$$\Rightarrow \quad V' = V/3. \qquad \textbf{(A1)}$$

Newton's Experimental Law: $eV = V - V'$ **(M1 A1)**

$$\Rightarrow \quad e = \tfrac{2}{3}.$$ **(A1)**

---

2 Diagram (or explanation) showing contact point of hemisphere and inclined plane in same vertical line with centre of mass of hemisphere. **(M2)**
$\sin \theta = \tfrac{3}{8}, \Rightarrow \theta \approx 22°.$ **(M1 A1, A1)**

---

3 Hooke's Law: $T = \lambda x/l$, where λ is the modulus of elasticity of the spring, **(M1)**

$$\Rightarrow \quad \lambda = lT/x.$$ **(A1)**

Work required to extend spring to new position

$$= \frac{1}{2}\lambda\frac{(2x)^2}{l} - \frac{1}{2}\lambda\frac{x^2}{l} = 3Tx/2.$$ **(M2 A2)**

---

4 $KE = \tfrac{1}{2} \times 50 \times (50)^2 \text{ J} = 62\,500 \text{ J}.$ **(M1 A1)**
$PE = 50 \times 20 \times 10 \text{ J} = 10\,000 \text{ J}.$ **(M1 A1)**
$Power = (62\,500 + 10\,000) \text{ W} = 72 \cdot 5 \text{ kW}.$ **(M1 A1)**

---

5 Suppose cars are level again after time T from O. **(M1)**
In time T, C_1 moves through a distance $2uT$. **(A1)**
In time T, C_2 moves through a distance $\tfrac{1}{2}fT^2$. **(A1)**
Since these distances are equal and $T \neq 0$,

$$T = 4u/f.$$ **(M1 A1)**

Distance required $= 8u^2/f.$ **(A1)**

---

6 Take unit vectors $\mathbf{i}$ and $\mathbf{j}$ in the directions east and north respectively. Then

$$\mathbf{R} = \left[(40\cos 45°)\mathbf{i} + (35 - 40\sin 45°)\mathbf{j}\right] N.$$ **(M1 A1)**

$$|\mathbf{R}| = \sqrt{\left[(20\sqrt{2})^2 + (35 - 20\sqrt{2})^2\right]} N \approx 29 \text{ N}.$$ **(M1 A1)**

Direction of $\mathbf{R}$ is $\tan^{-1}\left(\dfrac{20\sqrt{2}}{35 - 20\sqrt{2}}\right) \approx 077°.$ **(M1 A1)**

---

7 The greatest value of x occurs when the man stands on the beam between B and C so that the force exerted by the beam on the support of A is zero. When

this occurs, we have

$$\curvearrowleft B, \quad \Rightarrow \quad 2W(x - 2l) = Wl/2 \quad \Rightarrow \quad x = 9l/4. \qquad \textbf{(M2 A1)}$$

The least value of x occurs when the man stands on the beam between A and B and as near to A as the restriction on the magnitude of the force exerted by the beam on the support at A allows. That is, the magnitude is $2W$.

$$\curvearrowleft B, \quad \Rightarrow \quad 2W(2l) = W(l/2) + 2W(2l - x) \quad \Rightarrow \quad x = l/4.$$
$$\textbf{(M2 A1)}$$

The man may stand on the beam at any position between these critical points, so $l \leqslant 4x \leqslant 9l$. $\qquad \textbf{(A1)}$

---

8 Applying Newton's Second Law to each particle in turn, we have:

For the heavier, $\qquad\qquad\qquad 2mg - T = 2mf, \qquad\qquad\qquad \textbf{(M1 A1)}$

for the lighter, $\qquad\qquad\qquad T - mg = mf, \qquad\qquad\qquad\quad \textbf{(M1 A1)}$

$$\Rightarrow \quad f = g/3. \qquad\qquad\qquad\qquad \textbf{(M1 A1)}$$

Time t taken by the heavier particle to reach the plane is given by

$$3a = \tfrac{1}{2}(g/3)t^2 \quad \Rightarrow \quad t = 3\sqrt{(2a/g)}. \qquad \textbf{(M1 A1)}$$

---

9 The line of action of the force exerted by the bowl on the particle passes through the centre of the sphere. If the angle made by this line of action with the vertical is θ, $\qquad\qquad\qquad\qquad\qquad\qquad\qquad\qquad \textbf{(M1)}$

$$\updownarrow, \quad \Rightarrow \quad R \cos \theta = mg \text{ and } \cos \theta = 12/13, \text{ so } R = 13mg/12,$$
$$\textbf{(M1 A1, A1)}$$

$$\leftrightarrow, \quad \Rightarrow \quad R \sin \theta = mV^2/(5a), \quad \Rightarrow \quad V = 5\sqrt{(ga/3)}/2.$$
$$\textbf{(M1 A1, M1 A1)}$$

---

10 At time t, the equation of motion of the particle is

$$mv \frac{\mathrm{d}v}{\mathrm{d}x} = -mk^2 x. \qquad \textbf{(M1 A1)}$$

Integrating, $\qquad\qquad v^2 = -k^2 x^2 + C. \qquad\qquad \textbf{(M1 A1)}$

Since $v = 0$ when $x = a$, we have

$$C = k^2 a^2 \quad \text{and} \quad v^2 = k^2(a^2 - x^2). \qquad \textbf{(A1)}$$

The motion is simple harmonic with period $2\pi/k$, the time that elapses before P returns to A for the first time. $\qquad\qquad\qquad\qquad\qquad\qquad \textbf{(A1)}$

Using $x = a \cos kt$, the time taken for P to reach the mid-point of OA from A is $\pi/(3k)$. $\qquad\qquad\qquad\qquad\qquad\qquad\qquad\qquad\qquad\qquad \textbf{(M1 A1)}$

---

11 Velocity of B relative to $A = \left[(4\mathbf{i} - \mathbf{j}) - (3\mathbf{i} - 2\mathbf{j})\right]$ m s^{-1} (M2)

$$= (\mathbf{i} + \mathbf{j}) \text{ m s}^{-1}.$$ (A1)

If the particles collide at time t s, then the path of B relative to A passes through the point with position vector $(4\mathbf{i} + 4\mathbf{j})$ m.

That is, at some value of t, say, if it uniquely exists,

$$-4\sqrt{2}\mathbf{i} + p\mathbf{j} + t(\mathbf{i} + \mathbf{j}) = 4\mathbf{i} + 4\mathbf{j}.$$ (M2)

Equating $\mathbf{i}$ coefficients $\qquad \Rightarrow \quad t = 4 + 4\sqrt{2}.$

(M1 A1)

Equating $\mathbf{j}$ coefficients $\qquad \Rightarrow \quad p = -4\sqrt{2},$

and is the required value of p. (M1 A1)

---

12 Resolve the initial velocity of the particle into $V \cos \theta$ horizontally and $V \sin \theta$ vertically. (M1)

Time of flight from O to A is $(2V \sin \theta)/g$. (M1 A1)

The range OA is of length $(2V^2 \sin \theta \cos \theta)/g$. (M1 A1)

The greatest height achieved by the particle is $(V^2 \sin^2 \theta)/(2g)$, (M1 A1)

$$\Rightarrow \quad (V^2 \sin^2 \theta)/(2g) = (2V^2 \sin \theta \cos \theta)/g \quad \Rightarrow \quad \tan \theta = 4.$$

(M1 A1)

---

13 Newton's Second Law gives

$$2\frac{d\mathbf{v}}{dt} = 4\,e^{2t}\mathbf{i} + 16\,e^{-2t}\mathbf{j}.$$ (M1 A1)

Integrating, $\qquad \mathbf{v} = e^{2t}\mathbf{i} - 4\,e^{-2t}\mathbf{j} + \mathbf{c}.$ (M1 A1)

Initially, $\mathbf{v} = 2\mathbf{i} + 3\mathbf{j}$ when $t = 0$ and so $\mathbf{c} = \mathbf{i} + 7\mathbf{j}$. (M1 A1)

$t = \ln 2$, i.e. $e^t = 2$ and the speed of the particle at this instant is

$$|4\mathbf{i} - \mathbf{j} + \mathbf{i} + 7\mathbf{j}| \text{ m s}^{-1} = \sqrt{61} \text{ m s}^{-1}.$$ (M1 A2)

---

14 The event required can take place in two ways:

EITHER (green, black, black) OR (black, green, black).

(M2)

The probability of this event occurring $= \frac{2}{3} \times \frac{1}{3} \times \frac{1}{3} + \frac{1}{3} \times \frac{2}{3} \times \frac{1}{3}$ (M1)

$$= \frac{4}{27}.$$ (A1)

(a) P(B is selected *and* a green ball is drawn) $= \frac{1}{3} \times \frac{3}{8} = \frac{1}{8}$. (M1 A1)

(b) P(a green ball is selected) $= \frac{1}{3} \times \frac{4}{6} + \frac{1}{3} \times \frac{3}{8} + \frac{1}{3} \times \frac{2}{3}$ (M2 A1)

$$= \frac{41}{72}.$$ (A1)

7.6 Paper 10 (complete paper with choice)

Time allowed: $2\frac{1}{2}$ hours

Answer ALL questions in Section I and four questions in Section II. When required, take the magnitude of the acceleration due to gravity to be $10\,\mathrm{m\,s}^{-2}$.

SECTION I

Answer all questions in this section.

1 The position vector, relative to the origin O, of a particle P of mass m, moving in a horizontal plane in which $\mathbf{i}$ and $\mathbf{j}$ are perpendicular unit vectors, is given at time t by $\mathbf{r} = (a\cos kt)\mathbf{i} + (2a\sin 2kt)\mathbf{j}$, where a and k are positive constants. Find the speed of P at the instant when $t = \pi/(6k)$.

(5 marks)

2 A particle A, of mass $2m$ and moving with speed $4u$, strikes directly a particle B of mass $3m$ which is at rest. Immediately after the impact, B moves with speed $2u$. Calculate the speed of A after the impact and the coefficient of restitution between A and B.

(5 marks)

3 A conical pendulum consists of a light inextensible string, of length a, together with a bob of mass m attached to its free end. The bob describes a horizontal circle with constant angular speed ω and the string makes an angle θ with the horizontal. Find an equation relating ω^2, g, a and θ.

(6 marks)

4 Two particles A and B, of masses $0.3\,\mathrm{kg}$ and $0.4\,\mathrm{kg}$ respectively, are connected by a light inextensible string passing over a fixed light smooth pulley. The particles are released from rest with the string taut and the hanging parts vertical. Calculate, in N, the tension in the string and find, in J, the kinetic energy of the system after it has been in motion for $1.5\,\mathrm{s}$.

(6 marks)

5 Forces $3\mathbf{i}$ N, $(\mathbf{i} + 2\mathbf{j})$ N and $(2\mathbf{i} + 2\mathbf{j})$ N act at the points A, B and C respectively. The position vectors of A, B and C relative to an origin O are $2\mathbf{j}$ m, $2\mathbf{i}$ m and $(3\mathbf{i} + 2\mathbf{j})$ m respectively. Show that the line of action of the resultant of these forces passes through O.

(6 marks)

6 A uniform symmetrical composite body consists of a right circular cone, of base radius r and height h, and a hemisphere of radius r. The base of the cone and the plane face of the hemisphere coincide and are stuck firmly together. Given that the centre of mass of the composite body is situated at the common centre of the plane faces of the cone and hemisphere, express h in terms of r. Hence find, to the nearest degree, the vertical angle of the cone.

(6 marks)

7 A particle moves in a straight line so that its speed is inversely proportional to $(t + 2)$, where t is the time in seconds for which it has been moving. After 3 s, the particle has retardation of magnitude 2 m s^{-2}. Leaving your answers in terms of natural logarithms, calculate the distance moved by the particle in the first three seconds of its motion.

(7 marks)

8 Three boys and three girls sit in a row of six seats. Find the probability that

(a) the three girls sit together,
(b) the boys and girls sit in alternate seats,
(c) a particular boy and a particular girl sit next to each other.

(7 marks)

SECTION II

Answer FOUR questions in this section.

9 A particle P moves in a straight line Ox so that at time t its acceleration in the direction of increasing x is $U\omega^2 t\, e^{\omega t}$, where ω and U are positive constants. Given that P starts from rest at O when $t = 0$, find, in terms of U, ω and t, its distance from O at time t.

Find also expressions for the speed of P and its distance from O when $t = 1/\omega$.

(13 marks)

10 A particle P, of mass m, is connected by a light elastic spring, of natural length l and modulus $mg/2$, to a fixed point O on a smooth horizontal table. The particle lies on the table with $OP = l$ and it is given an impulse of magnitude $m\sqrt{(lg)}$ in the direction $\overrightarrow{OP}$. Find an expression for the maximum extension of the spring.

Find also the time that elapses before P first comes to instantaneous rest.

(13 marks)

---

11 A small boat, moving at 8 km h^{-1} relative to the water, travels directly from a point X to a point Y which is 10 km away and whose bearing from X is 150°. The boat then travels directly to a point Z, 10 km due west of Y, at the same speed relative to the water. Given that a current of constant speed 4 km h^{-1} is flowing throughout the region from north to south, find the two courses which the boat must set to reach Y and then Z.

Find also, to the nearest minute, the time taken to complete the total two stage journey.

(13 marks)

---

12 A light rectangular plate $ABCD$ has $AB = CD = 3a$ and $BC = DA = 2a$. The plate is placed on a smooth horizontal plane and lies flat on this plane. Horizontal forces of magnitudes $4P, 6P, 9P$ and $6P$ are applied to the plate and act along the lines $\overrightarrow{AB}, \overrightarrow{BC}, \overrightarrow{CD}$ and $\overrightarrow{AD}$ respectively. A fifth force acts in the line ZB, where Z lies in CD, and prevents the plate from moving. Calculate the magnitude of this fifth force and the distance DZ.

If this fifth force had acted through A, instead of in the line ZB, find the moment of the couple produced by the five forces.

(13 marks)

---

13 The end A of a uniform rod AB, of length $2a$ and mass M, is smoothly hinged at a fixed point to a vertical wall. A light string has one end tied at B and the other end is tied to a fixed point C in the wall, the point C being vertically above A and $AC = a$. Given that the rod and the string are of the same length, show that the magnitude of the tension in the string is Mg.

Calculate the magnitude and the direction of the force exerted by the hinge on the rod.

Given further that the string is elastic and of natural length a, find, in terms of M and g, the modulus of elasticity of the string.

When a particle of mass kM is attached to the rod at B, the system rests in equilibrium with AB horizontal. Find the value of k.

(13 marks)

---

14 A car, of mass 800 kg, is moving along a straight level road against constant resistive forces of total magnitude 700 N. At a particular instant the speed of the car is 15 m s^{-1} and the acceleration of the car is of magnitude $\frac{1}{2}$ m s^{-2}. Calculate, in kW, the rate at which the engine of the car is working.

The car ascends a hill, inclined at θ to the horizontal, with the engine working at the same rate and against the same constant resistive forces. Given that the greatest speed of the car up this hill is 9 m s^{-1}, calculate θ to the nearest degree.

At a specific instant when the car is climbing this hill with speed 9 m s^{-1}, the power output of the engine of the car is increased by 5 kW. Find, in m s^{-2}, the instantaneous acceleration of the car resulting from this power increase.

(13 marks)

8 Revision methods and short-question tests: pure mathematics

8.1 The topic method of revision

In order to undertake this approach you need a copy of your syllabus and at least three sets of past examination papers which have been set by your board in recent years.

Using cards or separate sheets of file paper make a list of topics from your syllabus which can be matched with several questions from your set of past papers. Obtain solutions to these questions concentrating your attention on one topic at a time. In the quest for solutions you should first of all try to do the questions on a topic without resorting to any external help. When you cannot do a question, look for a similar one in your textbooks or notes. If this approach fails, then ask your teacher for help or discuss the problem with one of your friends. You must be persistent until the topic is mastered. Finally write up the solutions to the questions on the topic in a neat and tidy fashion so that you can refer to them in your final days of revision. Each topic should receive this detailed, painstaking approach and you will find that your confidence will grow quite dramatically as you work through your list.

8.2 Working examination papers to time

Many students try to work whole papers to time before they are really ready and their morale and confidence are often sapped away completely.

There is no need for this to happen. The successful completion of the topic approach explained above is a prerequisite for success in tackling whole papers to time. Short questions, where the content demands are constantly changing every few minutes, are particularly demoralizing unless you are really prepared for them. Like a good athlete you must get fit by stages and it is best to start this part of your training by careful staging and planning. Suppose you have a target in a pure mathematics paper of eight Section I questions (all to be attempted) and six longer Section II questions of which FOUR must be done and that two and a half hours is allowed for the whole paper. Let us assume that Section I is worth 48 per cent of the total mark and that each Section II question is worth 13 per cent of the total mark. When you are completely prepared, this would mean that, after allowing a few minutes to read the paper through and plan your choice of strong questions to start with, you have about seventy minutes for Section I and about nineteen minutes per question for Section II.

Now you are ready to start training but, remembering the comparison with athletic training, do not try to cover the whole course at your first session. Some will find that thirty-five minutes of concentration tackling just four Section I questions is enough to start with and for the longer questions in Section II try doing two out of three in forty minutes. Your second stage of training could double the time and double the demands by trying a whole Section I or four questions out of six in Section II. In each of these stages, you will need to be very critical of yourself in terms of performance and also be very determined to stick to your planned schedule. Try to arrange that regular assessment of your work can be made by your teacher, so that you can tackle each stage with growing confidence and assurance. The final stage of your build-up is reached when you feel that you are ready to tackle complete papers and often this is best done during the Easter holiday, when you can plan to set aside several sessions of two and a half hours without disturbance. If you have taken a set of mock papers earlier in the year, it is a good confidence builder to tackle these again at this stage.

The following examples and tests are given to provide you with examples of the type of practice your revision should contain. They do not form a complete course in themselves but are just an indication of the types of targets you need to set yourself.

8.3 Formula booklets
Often in examinations, except those using multiple-choice components, candidates are issued with a list of formulae. If your board follows this practice, you should use this list throughout these stages of your revision programme. This use will get you completely familiar not only with the formulae given, but also where to find them quickly. If your school or centre is unable to supply you with this list or with a copy of your syllabus and past papers, all of these may be purchased directly from your examination board.

8.4 Revision notes on vectors
In both pure and applied mathematics the topic which is often most feared by students is vectors. Because of this, we have given a set of revision notes on this topic which cover all the basic facts. These notes will also serve as an illustration to the student of the way in which a set of revision notes should be prepared.

These notes give the basic facts you need to know about vectors. The proofs of various pieces of bookwork which are sometimes required in examinations can be found in standard text books.

A *vector* quantity has magnitude and direction, e.g. velocity, force, displacement.

A *scalar* quantity has magnitude only, e.g. mass, time, energy.

A vector can be represented geometrically by a directed line segment. The magnitude of the vector is then the length of the line segment and its direction is the direction of the line segment with respect to some standard reference such as a fixed origin and set of coordinate axes.

Be careful always to write your vectors in the proper way, i.e. arrow notation or underlining, as the printed boldface type is not available to you, e.g. $\overrightarrow{AB}$ or $\overline{AB}$ or $\underline{AB}$ and $\underline{a}, \underline{i}$ etc. *Never* omit the underlining as, if you do so, you are liable to make your work meaningless. The magnitude of $\underline{a}$ is denoted by $|\underline{a}|$ or a.

A *unit* vector has magnitude 1.

The *null* vector **(0)** has magnitude zero.

Equal vectors

Two vectors are equal if and only if they are of equal magnitude and in the same direction.

Multiplication by a scalar

Given that λ is a positive scalar, then $\lambda \mathbf{a}$ is a vector in the direction of **a** and of magnitude $\lambda |\mathbf{a}|$. When λ is negative, $\lambda \mathbf{a}$ is in the direction opposite to **a** and is of magnitude $|\lambda||\mathbf{a}|$. The special case when $\lambda = -1$ gives us that $-\mathbf{a}$ is equal in magnitude to **a** but opposite in direction.

Addition of vectors

The basis of this is the *vector triangle*.

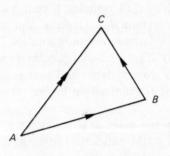

$$\overrightarrow{AC} = \overrightarrow{AB} + \overrightarrow{BC}.$$

Looked at in terms of simple displacements, the displacement from A to C is equivalent to the displacement from A to B followed by the displacement from B to C. Once you have mastered this concept then there is no need ever to be troubled by signs; the arrows in the diagram display exactly what you are doing.

Then, by using ordinary algebra, which can be proved to be valid for this purpose, we say

$$\overrightarrow{AB} = \overrightarrow{AC} - \overrightarrow{BC}.$$

The ratio theorem

A useful result in geometrical examples is the following:

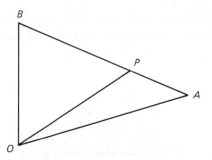

If the point P is such that $AP/PB = \lambda/\mu$, then
$$(\lambda + \mu)\overrightarrow{OP} = \mu\overrightarrow{OA} + \lambda\overrightarrow{OB}.$$

We note that, here, the *position* of the vectors is of some importance, all being referred to the point O, which we might define as the origin. These are then called *position vectors* as opposed to free vectors in which the position is not important.

An important special case of the ratio theorem occurs when $\lambda = \mu = 1$,

i.e. $\qquad 2\overrightarrow{OP} = \overrightarrow{OA} + \overrightarrow{OB} \qquad$ or $\qquad \overrightarrow{OP} = \tfrac{1}{2}(\overrightarrow{OA} + \overrightarrow{OB}),$

which defines the position vector of the mid-point of AB.

In statics this gives the position vector of the centre of mass of equal particles at A and B and it generalizes to the following:

The centre of mass G, of particles with masses $m_1, m_2, \ldots, m_n$, placed at points with position vectors $\mathbf{r}_1, \mathbf{r}_2, \ldots, \mathbf{r}_n$ respectively, is defined by

$$\overrightarrow{OG} = \frac{\displaystyle\sum_{i=1}^{n} m_i\mathbf{r}_i}{\displaystyle\sum_{i=1}^{n} m_i},$$

where O is the origin.

Addition of position vectors

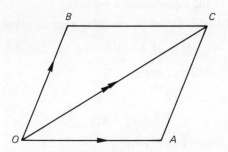

To find $\overrightarrow{OA} + \overrightarrow{OB}$, complete parallelogram $OACB$. Then
$$\overrightarrow{OA} + \overrightarrow{OB} = \overrightarrow{OA} + \overrightarrow{AC} = \overrightarrow{OC}.$$

Hence $\overrightarrow{OC}$ represents the sum of $\overrightarrow{OA}$ and $\overrightarrow{OB}$ in magnitude, direction and position. To find the *difference* of position vectors, complete parallelogram *OBAD*.

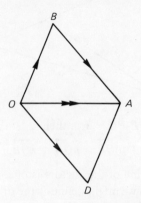

$$\overrightarrow{OA} = \overrightarrow{OB} + \overrightarrow{BA} = \overrightarrow{OB} + \overrightarrow{OD}$$
$$\Rightarrow \quad \overrightarrow{OD} = \overrightarrow{OA} - \overrightarrow{OB}$$

and so $\overrightarrow{OD}$ represents the *difference* of $\overrightarrow{OA}$ and $\overrightarrow{OB}$ in magnitude, direction and position.

Use of coordinate axes

Unit vectors in the direction of the coordinate axes Ox, Oy and Oz are defined as **i**, **j** and **k**, respectively. Consider vector $\overrightarrow{OA} = \mathbf{a}$, where O is fixed origin and (a_1, a_2, a_3) are the coordinates of A with respect to axes Ox, Oy, Oz.

Then
$$\overrightarrow{OA} = a_1\mathbf{i} + a_2\mathbf{j} + a_3\mathbf{k},$$

often written more shortly in the coordinate form (a_1, a_2, a_3) or as a column

vector $\begin{pmatrix} a_1 \\ a_2 \\ a_3 \end{pmatrix}$. (This expression of a vector in terms of its components is used in more advanced work as the *definition* of a vector.)

The *magnitude* of $\overrightarrow{OA}$, also called its modulus, may be written

$$|\overrightarrow{OA}| \text{ or } |\mathbf{a}| = \sqrt{(a_1^2 + a_2^2 + a_3^2)}, \qquad \text{i.e. the distance } OA.$$

The *unit* vector in direction of $\overrightarrow{OA}$ is

$$\frac{a_1\mathbf{i} + a_2\mathbf{j} + a_3\mathbf{k}}{\sqrt{(a_1^2 + a_2^2 + a_3^2)}} = \hat{\mathbf{a}}.$$

The sum of vectors in coordinate form
$$\overrightarrow{OA} = a_1\mathbf{i} + a_2\mathbf{j} + a_3\mathbf{k}, \qquad \overrightarrow{OB} = b_1\mathbf{i} + b_2\mathbf{j} + b_3\mathbf{k},$$
$$\Rightarrow \quad \overrightarrow{OA} + \overrightarrow{OB} = (a_1 + b_1)\mathbf{i} + (a_2 + b_2)\mathbf{j} + (a_3 + b_3)\mathbf{k}.$$

Scalar or dot product

Given two vectors **a** and **b** with an angle θ between them, the scalar product **a . b** is defined by

$$\mathbf{a . b} = ab \cos \theta.$$

When $\theta = 0$, $\cos \theta = 1$ and $\mathbf{a . b} = ab$.
When $\theta = \pi/2$, $\cos \theta = 0$ and $\mathbf{a . b} = 0$.

Hence

$$\mathbf{i . i} = \mathbf{j . j} = \mathbf{k . k} = 1,$$

$$\mathbf{i . j} = \mathbf{j . k} = \mathbf{k . i} = 0,$$

Therefore

$$\mathbf{a . b} = a_1 b_1 + a_2 b_2 + a_3 b_3.$$

Differentiation and integration of vectors

Given that $\mathbf{r} = x\mathbf{i} + y\mathbf{j} + z\mathbf{k}$, where x, y, z are functions of the scalar variable t, then

$$\frac{\mathrm{d}}{\mathrm{d}t}(\mathbf{r}) = \mathbf{i}\frac{\mathrm{d}x}{\mathrm{d}t} + \mathbf{j}\frac{\mathrm{d}y}{\mathrm{d}t} + \mathbf{k}\frac{\mathrm{d}z}{\mathrm{d}t}$$

or, using the 'dot' notation,

$$\dot{\mathbf{r}} = \dot{x}\mathbf{i} + \dot{y}\mathbf{j} + \dot{z}\mathbf{k}.$$

This is particularly useful for finding velocity and acceleration from a position vector.

By *definition*,

$$\text{velocity } \mathbf{v} = \dot{\mathbf{r}},$$

$$\text{acceleration } \mathbf{f} = \ddot{\mathbf{r}} = \ddot{x}\mathbf{i} + \ddot{y}\mathbf{j} + \ddot{z}\mathbf{k}.$$

When the speed only is required $|\mathbf{v}| = v = |\dot{\mathbf{r}}| = \sqrt{(\dot{x}^2 + \dot{y}^2 + \dot{z}^2)}$, and the magnitude of the acceleration $= |\ddot{\mathbf{r}}| = \sqrt{(\ddot{x}^2 + \ddot{y}^2 + \ddot{z}^2)}$.

Integration is the reverse of differentiation, but be careful never to omit the constant of integration which, in this case, is a vector. Thus

$$\mathbf{v} = \dot{x}\mathbf{i} + \dot{y}\mathbf{j} + \dot{z}\mathbf{k},$$

$$\Rightarrow \quad \mathbf{r} = \int \mathbf{v}\, \mathrm{d}t = x\mathbf{i} + y\mathbf{j} + z\mathbf{k} + \mathbf{c},$$

where **c** is the constant of integration and can be found by the substitution of particular (usually initial) values of **r** and t.

Geometrical aspects

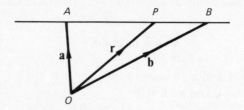

Given that the position vectors of A and B with respect to O are $\mathbf{a}$ and $\mathbf{b}$ respectively, then an equation of AB is

$$\mathbf{r} = \mathbf{a} + t(\mathbf{b} - \mathbf{a}), \qquad \text{where } t \text{ is a scalar parameter,}$$

or $\qquad \mathbf{r} = \mathbf{a} + t\mathbf{c}, \qquad$ where $\mathbf{c}$ is in the direction $\overrightarrow{AB}$.

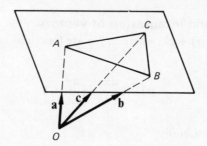

Given that the position vectors of A, B and C are $\mathbf{a}$, $\mathbf{b}$ and $\mathbf{c}$ respectively, then an equation of the plane ABC is

$$\mathbf{r} = \mathbf{a} + (\mathbf{b} - \mathbf{a})\lambda + (\mathbf{c} - \mathbf{a})\mu,$$

or $\qquad \mathbf{r} = \mathbf{a} + \lambda\mathbf{d} + \mu\mathbf{e},$

where $\mathbf{d}$ and $\mathbf{e}$ are in the directions of $\overrightarrow{AB}$ and $\overrightarrow{AC}$ respectively.

Normal form of the equation of a plane

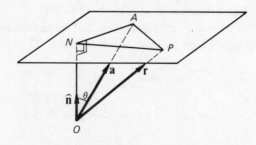

For an equation of the plane through the point A (where $\overrightarrow{OA} = \mathbf{a}$) and with $\hat{\mathbf{n}}$ the unit vector in the direction of the normal ON from O to the plane,

$$(\mathbf{r} - \mathbf{a}) \cdot \hat{\mathbf{n}} = 0 \qquad \text{(as } \hat{\mathbf{n}} \text{ is perpendicular to } \overrightarrow{AP})$$

$$\Rightarrow \quad \mathbf{r} \cdot \hat{\mathbf{n}} = \mathbf{a} \cdot \hat{\mathbf{n}} = p, \qquad \text{where } p = ON.$$

The equation $\mathbf{r} \cdot \hat{\mathbf{n}} = p$ is a normal or perpendicular form of the equation of the plane.

Link-up with 3-dimensional coordinate geometry

Taking $\mathbf{r} = x\mathbf{i} + y\mathbf{j} + z\mathbf{k}$ and letting the direction cosines of the normal from O to the plane be (n_1, n_2, n_3), i.e.

$$\hat{\mathbf{n}} = n_1\mathbf{i} + n_2\mathbf{j} + n_3\mathbf{k},$$

the equation $\mathbf{r} \cdot \hat{\mathbf{n}} = p$ becomes

$$n_1 x + n_2 y + n_3 z = p,$$

which is the equation of a plane in its simplest form.

Note that here n_1, n_2, n_3 are *actual* direction cosines, i.e. $\sqrt{(n_1^2 + n_2^2 + n_3^2)} = 1$, and p is the distance from O to the plane. If the equation is given in the form $ax + by + cz = d$, where a, b, c are *not* the actual direction cosines, then it is easily converted by dividing by $\sqrt{(a^2 + b^2 + c^2)}$, so that the distance from O to the plane is $\dfrac{|d|}{\sqrt{(a^2 + b^2 + c^2)}}$.

8.5 Examples using the topic method of revision

Examples 1 Vectors
Consider the following three short questions on vectors:

1 Relative to an origin O, the points A, B and P have position vectors $\mathbf{a}$, $\mathbf{b}$ and $\frac{1}{4}\mathbf{a}$ respectively. The point R lies on AB so that $2\overrightarrow{AR} = \overrightarrow{RB}$. Giving your answers in terms of $\mathbf{a}$ and $\mathbf{b}$,

   (a) write down the position vector of R,
   (b) find the position vector of the point Q so that $BQPR$ is a parallelogram.
   (6 marks)

2 The particles A and B start to move simultaneously from the origin O and the point with position vector $(-10\mathbf{i} + 16\mathbf{j})$ m respectively. Particle A moves with constant velocity of magnitude $12 \cdot 5$ m s^{-1} in the direction of the vector $7\mathbf{i} + 24\mathbf{j}$ and particle B moves with constant velocity $(6\mathbf{i} + 8\mathbf{j})$ m s^{-1}. Prove that the particles collide and find the position vector of their point of collision.
   (6 marks)

3 Given that $\overrightarrow{OA} = (3\mathbf{i} + 4\mathbf{j} - 5\mathbf{k})$ and $\overrightarrow{OB} = (4\mathbf{i} + 2\mathbf{j} - 3\mathbf{k})$, where O is the origin, find

(a) an equation for the line AB in vector form,

(b) the position vector of the point C, on AB, such that OC is perpendicular to AB.

(7 marks)

Hints and solutions for Examples 1

1 Your first step is to sketch a diagram showing all given information and this is left for you to do.

(a)
$$\overrightarrow{OR} = \frac{2\overrightarrow{OA} + 1\overrightarrow{OB}}{3} = \tfrac{1}{3}(2\mathbf{a} + \mathbf{b}) \quad \text{(ratio theorem).} \qquad \textbf{(B1)}$$

(b) Let Q have position vector $\mathbf{q}$, i.e. $\overrightarrow{OQ} = \mathbf{q}$. Then

$$\overrightarrow{QB} = \overrightarrow{PR} \qquad \textbf{(M1)}$$

$$\Rightarrow \quad \overrightarrow{OB} - \overrightarrow{OQ} = \overrightarrow{OR} - \overrightarrow{OP} \qquad \textbf{(M1)}$$

$$\Rightarrow \quad \mathbf{b} - \mathbf{q} = \tfrac{2}{3}\mathbf{a} + \tfrac{1}{3}\mathbf{b} - \tfrac{1}{4}\mathbf{a} = \tfrac{5}{12}\mathbf{a} + \tfrac{1}{3}\mathbf{b} \qquad \textbf{(A2)}$$

$$\Rightarrow \quad \mathbf{q} = -\tfrac{5}{12}\mathbf{a} + \tfrac{2}{3}\mathbf{b}. \qquad \textbf{(A1 cao)}$$

---

2 Draw a sketch showing all information. **(B1)**

The unit vector along the path of A is $\dfrac{7\mathbf{i} + 24\mathbf{j}}{25}$. **(B1)**

So, A moves along the line $\mathbf{r} = \left(\dfrac{12\cdot 5}{25}\right)t(7\mathbf{i} + 24\mathbf{j})$, **(M1)**

where t s is the time since A started from O.

Similarly, B moves along the line given by

$$\mathbf{r} = (-10\mathbf{i} + 16\mathbf{j}) + t(6\mathbf{i} + 8\mathbf{j}). \qquad \textbf{(M1)}$$

When $t = 4$, the position vector of both A and B is $(14\mathbf{i} + 48\mathbf{j})$ m and they collide at this instant. **(A2)**

---

3 (a)
$$\overrightarrow{AB} = \overrightarrow{OB} - \overrightarrow{OA} = \mathbf{i} - 2\mathbf{j} + 2\mathbf{k}. \qquad \textbf{(B1)}$$

As t takes different values, the point P with position vector $\mathbf{r}$, where

$$\mathbf{r} = 3\mathbf{i} + 4\mathbf{j} - 5\mathbf{k} + t(\mathbf{i} - 2\mathbf{j} + 2\mathbf{k}), \qquad \textbf{(M1 A1)}$$

always lies on the line AB and this is an equation for AB in vector form.

(b) The point C on AB such that OC is perpendicular to AB is given by the value of t when the scalar product $\overrightarrow{AB} \cdot \overrightarrow{OC} = 0$ **(M1)**

$$\Rightarrow \quad 1(3 + t) - 2(4 - 2t) + 2(2t - 5) = 0 \qquad \textbf{(A1 FT)}$$

$$\Rightarrow \quad t = \tfrac{5}{3} \qquad \textbf{(A1 cao)}$$

$$\Rightarrow \quad \overrightarrow{OC} = 3\mathbf{i} + 4\mathbf{j} - 5\mathbf{k} + \tfrac{5}{3}(\mathbf{i} - 2\mathbf{j} + 2\mathbf{k}) = \tfrac{14}{3}\mathbf{i} + \tfrac{2}{3}\mathbf{j} - \tfrac{5}{3}\mathbf{k}. \qquad \textbf{(A1 cao)}$$

.Further examples on vectors are to be found in Short-question test 3 (p. 163) and in § 9.2.

Examples 2   Trigonometric equations

Consider the following four short questions which we will suppose you have selected from your set of past papers:

1 Solve, for $0 \leqslant \theta \leqslant 360$,

$$4 \cos \theta° - 3 \sin \theta° = 2. \qquad \textbf{(6 marks)}$$

---

2 Solve, for $0 \leqslant \theta \leqslant 360$,

$$\cos \theta° + 3 \sin \theta° = -\sec \theta°. \qquad \textbf{(6 marks)}$$

---

3 Write down expressions for $\cos 2\theta$ and $\sin 2\theta$ in terms of t, where $t = \tan \theta$. Hence, or otherwise, solve, for $0 \leqslant \theta \leqslant 360$,

$$5 \cos 2\theta° - 2 \sin 2\theta° = 2. \qquad \textbf{(6 marks)}$$

---

4 Giving your answers in terms of π, solve for $0 \leqslant \theta \leqslant \pi$,

$$\cos \theta + \cos 3\theta + \cos 5\theta = 0. \qquad \textbf{(6 marks)}$$

Hints and solutions for Examples 2

The following will help you build up your revision sheet for this topic.

1 The expression $4 \cos \theta° - 3 \sin \theta°$ can be written in the form $R \cos (\theta° + A°)$, where $R > 0$ and $0 < A < 90$. Use the identity

$$4 \cos \theta - 3 \sin \theta \equiv R \cos \theta \cos A - R \sin \theta \sin A, \qquad \textbf{(M1)}$$

to get $R = 5$, $\cos A° = \tfrac{4}{5}$ and $\sin A° = \tfrac{3}{5}$.   **(A1 A1)**

    These values of $\cos A°$ *and* $\sin A°$ taken together give $A = 36·87$ (to two decimal places).   **(A1 FT)**

The original equation is therefore equivalent to the simpler equation $5 \cos(\theta° + A°) = 2$, where $A = 36\cdot87$ (to two decimal places).

The answers are $29\cdot6$ and $256\cdot7$ (to one decimal place).   **(A1 A1 cao)**

This type of question is often set and both this method and the alternative method shown in question 3 should be mastered thoroughly. You should learn both methods because a particular question may require you to use just one of these approaches and exclude the other.

---

2 Although this equation may look similar to question 1, the approach in solving it is quite different.

Multiplying by $\sec \theta°$ gives

$$1 + 3 \tan \theta° = -\sec^2 \theta°. \qquad \textbf{(M1)}$$

Also, $\sec^2 \theta = 1 + \tan^2 \theta$, and the equation may be written as

$$\tan^2 \theta° + 3 \tan \theta° + 2 = 0. \qquad \textbf{(M1 A1)}$$

Hence $\tan \theta° = -1$ or -2 **(A1 cao)**
and the solutions are

$$135, \quad 315, \quad 116\cdot6 \quad \text{and} \quad 296\cdot6,$$

the last two being given to one decimal place.   **(A2, 1 or 0)**

---

3 When a question states 'write down' you should be able to do this without working, either using your formula sheet or from memory!

$$\cos 2\theta = \frac{1 - t^2}{1 + t^2} \quad \text{and} \quad \sin 2\theta = \frac{2t}{1 + t^2}, \quad \text{where } t = \tan \theta.$$

(A1 A1 cao)

Using these identities for $\cos 2\theta$ and $\sin 2\theta$ in the given equation gives

$$7t^2 + 4t - 3 = 0 \qquad \textbf{(A1)}$$

and hence $\tan \theta° = -1 \quad \text{or} \quad \tfrac{3}{7},$ **(A1)**

and the solutions are

$$135, \quad 315, \quad 23\cdot2 \quad \text{and} \quad 203\cdot2, \qquad \textbf{(A2, 1 or 0)}$$

the last two being given to one decimal place.

---

4 The key to this solution depends on being able to rewrite the left-hand side of the equation in terms of factors.

Note that

$$\cos 5\theta + \cos \theta = 2 \cos 3\theta \cos 2\theta$$

from use of the identity   (M1)

$$\cos A + \cos B = 2 \cos \left(\frac{A + B}{2}\right) \cos \left(\frac{A - B}{2}\right),$$

which is likely to be given on your formula sheet.

The equation may now be written in the form

$$\cos 3\theta(2 \cos 2\theta + 1) = 0.$$ (A1)

Either $\cos 3\theta = 0$ or $\cos 2\theta = -\tfrac{1}{2}$,   (A1 A1 FT)

and the solutions are

$$\pi/6, \quad \pi/3, \quad \pi/2, \quad 2\pi/3, \quad 5\pi/6.$$ (A2, 1 or 0)

Note that we are required to give solutions in terms of π.

8.6   Short-question tests

Short-question test 1

Time allowed: 35 minutes

1   Prove that the tangent at $P(1/k, e)$ to the curve $y = e^{kx}$ passes through the origin O.

Find, in terms of e and k, the area of the finite region bounded by the curve, the line OP and the y-axis.

(6 marks)

---

2   Given that x is so small that x^3 and higher powers of x may be neglected, prove that

(a) $2\sqrt{(1 + 2x)} = 2 + 2x - x^2$,

(b) $(\sin x + \tan x)^2 = 8 - 8 \cos x$.

(6 marks)

---

3 The first, second and third terms of an arithmetic series are 1, p and q respectively. The first, second and third terms of a geometric series are 1, q and p respectively. Given that r, the common ratio of the geometric series is such that $|r| < 1$, find

(a) the values of p and q,

(b) the sum to infinity of the geometric series.

(6 marks)

---

4 Given that $f(x) = x^4 - 4x^3 + x^2 + 8x - 6$,

(a) show that $f(\pm\sqrt{2}) = 0$,

(b) find the two real, distinct, integral roots of the equation $f(x) = 0$.

(6 marks)

Hints and solutions for short-question test 1

1 Tangent equation $y = k\,e\,x$, which passes through O. **(M1 A1)**
Draw a sketch to show region. **(B1)**
Area of region = area under curve − area of triangle. **(M1)**
$$= \frac{e-2}{2k}.$$ **(M1 A1)**

---

2 (a) Binomial series. **(M1 A2, 1 or 0)**
(b) Use the approximations $\sin x \approx x$, $\tan x \approx x$, $\cos x \approx 1 - \frac{1}{2}x^2$.
(M1 M1 M1)

---

3 For the arithmetic series $2p = 1 + q$. **(M1)**
For the geometric series $q^2 = p$. **(M1)**
$p = \frac{1}{4}$, $q = -\frac{1}{2}$. **(A1 A1 cao)**
Sum of geometric series to infinity $= \frac{2}{3}$. **(M1 A1 FT)**

---

4 $(x - \sqrt{2})(x + \sqrt{2})$ are factors of $f(x) \Rightarrow x^2 - 2$ factor of $f(x)$.
(M1 A1)
$f(x) = (x^2 - 2)(x^2 - 4x + 3)$ (or use factor theorem). **(M1 A1)**
$f(x) = 0$ when $x = 1$ and 3. **(M1 A1)**

Short-question test 2 (curve sketching)

Time allowed: 50 minutes

1   Given that $f(x) \equiv x^2 - 6x + 12$, show that $f(x) > 0$ for all real values of x.
Using the same axes sketch the graphs of

(a) $y = f(x)$,   (b) $y = \dfrac{1}{f(x)}$.

(7 marks)

2   Using the same axes sketch the curves $y = \dfrac{1}{x - 2}$ and $y = \dfrac{x - 1}{x + 2}$, giving the
equations of the asymptotes.
Hence, or otherwise, find the set of values of x for which

$$\frac{1}{x - 2} \geqslant \frac{x - 1}{x + 2}.$$

(9 marks)

3   Show that if x is real the function $g(x)$, where $g(x) = \dfrac{3x + 2}{x(x - 2)}$, cannot take
values between $-\frac{1}{2}$ and $-4\frac{1}{2}$. Sketch the curve $y = g(x)$.

(9 marks)

4   Sketch the curve whose parametric equations are $x = t^2 + 1$, $y = 2t$.

(4 marks)

5   Find the turning points of the curve $y = \dfrac{(x - 1)^2}{4x^2 - 1}$. Sketch the curve and state
the equations of the asymptotes to the curve.

(9 marks)

Short-question test 3 (vectors)

Time allowed: 40 minutes

1   Given that $|\mathbf{a}| = 7, |\mathbf{b}| = 3$ and $|\mathbf{a} - \mathbf{b}| = 6$ find in degrees to one decimal place
the angle between the vectors $\mathbf{a}$ and $\mathbf{b}$.

(6 marks)

2 The vertices of a quadrilateral, whose internal angles are all less than $180°$, have position vectors $\mathbf{a}, \mathbf{b}, \mathbf{c}$ and $\mathbf{d}$ relative to a fixed origin O. Prove that the mid-points of the sides of the quadrilateral form the vertices of a parallelogram.

(6 marks)

---

3 Find a vector equation of the straight line joining the points $A(2, -1, 3)$ and $B(4, 3, -2)$. Determine the coordinates of the point P where this line meets the plane $x = 1$.

(7 marks)

---

4 Two lines have the vector equations $\mathbf{r} = \mathbf{k} + s(\mathbf{k} - \mathbf{i})$ and $\mathbf{r} = \mathbf{k} + t(\mathbf{i} + \mathbf{j})$. Given that they intersect at the point A find the position vector of A.

Find a vector perpendicular to both the given lines and hence, or otherwise, obtain a vector equation of the plane containing the two lines.

(10 marks)

Short-question test 4 (quadratic function, quadratic equation)

Time allowed: 35 minutes

1 Given that $f(x) \equiv x^2 + (k + 2)x + 2k$, show that

(a) the roots of $f(x) = 0$ are real for all real k,
(b) the roots of the equation $4f(x - k) + (2 - k)^2 = 0$ are real.

(7 marks)

---

2 Given that $y = \dfrac{x^2 + 4x + \lambda}{2x - 3}$ and x is real, find the greatest value of λ for which y can take all real values.

(7 marks)

---

3 If x can take values in the range $-5 \leqslant x \leqslant 6$, find the greatest and least values of $3x^2 - 7x - 4$.

(6 marks)

4 If α and β are roots of the equation $ax^2 + bx + c = 0$, find equations whose roots are

(a) $\dfrac{1}{\alpha}$ and $\dfrac{1}{\beta}$, (b) $\dfrac{\alpha^2}{\beta}$ and $\dfrac{\beta^2}{\alpha}$.

(6 marks)

9 Revision methods and short-question tests: applied mathematics

9.1 The topic method of revision

The ideas of §§ 8.1, 8.2 and 8.3 are relevant in applied mathematics also, and should be reread before the remainder of this chapter.

9.2 Worked examples with mark schemes

Examples 1 Vectors

Consider the following three short questions on vectors.

1 A particle starts from rest at the origin O and moves in a horizontal plane with constant acceleration $(2\mathbf{i} - \mathbf{j})$ m s^{-2}. The initial velocity of the particle is $(3\mathbf{i} + 2\mathbf{j})$ m s^{-1}. Find the position vector $\mathbf{r}$ m, of the particle at time t seconds after leaving O.

(5 marks)

---

2 A particle P starts at the origin O and at time t seconds the velocity of the particle is $\left[3t^2\,\mathbf{i} + (\cos t)\,\mathbf{j} + e^t\,\mathbf{k}\right]$ m s^{-1}. Find (a) the speed of the particle when $t = \pi$, (b) the position vector of the particle at time $t = 2\pi$, (c) the acceleration vector of the particle at time t.

(9 marks)

---

3 Referred to origin O, the position vectors of the points A and B are $(4\mathbf{i} + 3\mathbf{j})$ m and $(-12\mathbf{i}$ and $5\mathbf{j})$ m respectively. Forces $\mathbf{F}_1$ and $\mathbf{F}_2$, of magnitudes 15 N and 26 N respectively, act along $\overrightarrow{OA}$ and $\overrightarrow{OB}$. Calculate (a) the resultant $\mathbf{R}$ of $\mathbf{F}_1$ and $\mathbf{F}_2$, (b) the position vector of the point C in AB through which the line of action of this resultant passes, (c) the cosine of angle AOB.

(8 marks)

Hints and solutions for Examples 1

1 We have $\ddot{\mathbf{r}} = 2\mathbf{i} - \mathbf{j}$ and, integrating with respect to t,
$$\dot{\mathbf{r}} = 2t\mathbf{i} - t\mathbf{j} + \mathbf{u}, \text{ where } \mathbf{u} \text{ is a constant vector.} \quad \textbf{(M1)}$$

We are given that $\dot{\mathbf{r}} = 3\mathbf{i} + 2\mathbf{j}$ when $t = 0 \Rightarrow \mathbf{u} = 3\mathbf{i} + 2\mathbf{j}$.
This gives

$$\dot{\mathbf{r}} = (2t + 3)\mathbf{i} + (2 - t)\mathbf{j}. \qquad \textbf{(A1 cao)}$$

Integrating again with respect to t,

$$\mathbf{r} = (t^2 + 3t)\mathbf{i} + (2t - \tfrac{1}{2}t^2)\mathbf{j} + \mathbf{c}, \quad \text{where } \mathbf{c} \text{ is a constant vector.}$$
$$\textbf{(M1)}$$

We are given that $\mathbf{r} = \mathbf{0}$ when $t = 0 \Rightarrow \mathbf{c} = \mathbf{0}$.   **(A1 FT)**
The position vector of the particle at time t is

$$\left[(t^2 + 3t)\mathbf{i} + (2t - \tfrac{1}{2}t^2)\mathbf{j}\right] \text{ m}. \qquad \textbf{(A1 cao)}$$

---

2 Let the position vector of P be $\mathbf{r}$ m at time t seconds.

(a) Then $\qquad \dot{\mathbf{r}} = 3t^2\mathbf{i} + (\cos t)\mathbf{j} + e^t\,\mathbf{k}$.

When $t = \pi$,

$$\dot{\mathbf{r}} = 3\pi^2\mathbf{i} - \mathbf{j} + e^\pi\,\mathbf{k}. \qquad \textbf{(M1)}$$

The speed of P at this instant

$$= \sqrt{(9\pi^4 + 1 + e^{2\pi})} \text{ m s}^{-1} \approx 37{\cdot}6 \text{ m s}^{-1} \text{ (3 s.f.).} \qquad \textbf{(A1 cao)}$$

(b) Integrating the velocity equation with respect to time

$$\Rightarrow \quad \mathbf{r} = t^3\mathbf{i} + (\sin t)\mathbf{j} + e^t\,\mathbf{k} + \mathbf{c}, \qquad \textbf{(M1 A1 cao)}$$

where $\mathbf{c}$ is a constant vector.
Since $\mathbf{r} = \mathbf{0}$ when $t = 0$, we have $\mathbf{c} = -\mathbf{k}$.

$$\therefore \quad \mathbf{r} = t^3\mathbf{i} + (\sin t)\mathbf{j} + (e^t - 1)\mathbf{k}. \qquad \textbf{(M1 A1 FT)}$$

When $t = 2\pi$, the position vector of P is

$$\left[8\pi^3\mathbf{i} + (e^{2\pi} - 1)\mathbf{k}\right] \text{ m}. \qquad \textbf{(A1 cao)}$$

(c) By differentiating the velocity equation with respect to time we obtain

$$\ddot{\mathbf{r}} = 6t\mathbf{i} - (\sin t)\mathbf{j} + e^t\,\mathbf{k},$$

which gives the acceleration vector of P as

$$\left[6t\mathbf{i} - (\sin t)\mathbf{j} + e^t\,\mathbf{k}\right] \text{ m s}^{-1}. \qquad \textbf{(M1 A1 cao)}$$

---

3 We show all the given information in a diagram.

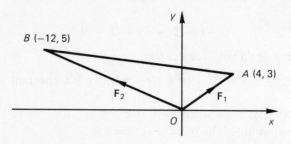

(a) Since the magnitude of $(4\mathbf{i} + 3\mathbf{j})$ is 5, the unit vector along $\overrightarrow{OA}$ is $(4\mathbf{i} + 3\mathbf{j})/5$.

Similarly, the unit vector along $\overrightarrow{OB}$ is $\frac{1}{13}(-12\mathbf{i} + 5\mathbf{j})$. **(M1)**

Since $|\mathbf{F}_1| = 15$,

$$\mathbf{F}_1 = 3(4\mathbf{i} + 3\mathbf{j}) \text{ N}$$

and, in the same way,

$$\mathbf{F}_2 = 2(-12\mathbf{i} + 5\mathbf{j}) \text{ N},$$

$$\mathbf{R} = \mathbf{F}_1 + \mathbf{F}_2 = (-12\mathbf{i} + 19\mathbf{j}) \text{ N}. \quad \textbf{(M1 A1)}$$

(b) Since the magnitude of $\mathbf{F}_1$ is 15 N and the force acts in the line OA, we may represent $\mathbf{F}_1$ by the vector $3\overrightarrow{OA}$ and for similar reasons we may represent the force $\mathbf{F}_2$ by the vector $2\overrightarrow{OB}$. **(M1)**

Using the ratio theorem we have $3\overrightarrow{OA} + 2\overrightarrow{OB} = 5\overrightarrow{OC}$,

$$\overrightarrow{OC} = \tfrac{3}{5}(4\mathbf{i} + 3\mathbf{j}) \text{ m} + \tfrac{2}{5}(-12\mathbf{i} + 5\mathbf{j}) \text{ m} = (-\tfrac{12}{5}\mathbf{i} + \tfrac{19}{5}\mathbf{j}) \text{ m}. \quad \textbf{(M1 A1)}$$

(c) Using the scalar product $\overrightarrow{OA} \cdot \overrightarrow{OB}$ in the form

$$\overrightarrow{OA} \cdot \overrightarrow{OB} = |\overrightarrow{OA}||\overrightarrow{OB}| \cos A\hat{O}B$$

$$\Rightarrow \quad (4\mathbf{i} + 3\mathbf{j}) \cdot (-12\mathbf{i} + 5\mathbf{j}) = |4\mathbf{i} + 3\mathbf{j}||-12\mathbf{i} + 5\mathbf{j}| \cos A\hat{O}B$$

$$\Rightarrow \quad \cos A\hat{O}B = \frac{-33}{5 \times 13} = -\frac{33}{65}. \quad \textbf{(M1 A1)}$$

Examples 2   Projectiles

Consider the following short questions which we will suppose you have selected from your set of past papers.

1   A particle P is projected from a point O and moves freely under gravity. The initial velocity of P is $(u\mathbf{i} + v\mathbf{j})$, where $\mathbf{i}$ is a unit vector directed horizontally and **j** is a unit vector vertically upwards. Find the greatest height of P above the horizontal plane through O. Find also the range on this plane.

(8 marks)

---

2 A particle P, projected from a point O on horizontal ground, moves freely under gravity and hits the ground again at the point A. Referred to O as origin, OA as x-axis and the upward vertical at O as y-axis, the equation of the path of P is $y = x - x^2/40$, where x and y are measured in metres. Taking the acceleration due to gravity to be 10 m s^{-2}, calculate

(a) the distance OA,   (b) the velocity of P at O.

(8 marks)

---

3 A particle P is projected with velocity of magnitude 65 m s^{-1} at an angle of elevation α, where $\tan \alpha = \frac{5}{12}$, from a point O at the top of a cliff, 70 m above the sea. The particle moves freely under gravity and strikes the sea at the point C. Taking the acceleration due to gravity to be 10 m s^{-2}, calculate

(a) the time taken by P to reach C from O,
(b) the horizontal distance between O and C,
(c) the velocity of P at C.

(11 marks)

Hints and solutions for Examples 2

1 We suppose that P attains the highest point of its path at H and that P meets the horizontal plane through O at the point A, as shown in the diagram. Let us suppose that, at time t after projection, P is at the point distance x from O horizontally and distance y from O vertically.

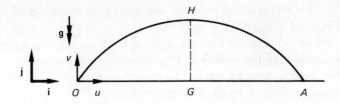

The acceleration of P at any instant is of magnitude g directed vertically downwards, so we can write

$$\ddot{x}\mathbf{i} + \ddot{y}\mathbf{j} = -g\mathbf{j}. \qquad \textbf{(M1)}$$

Integrating with respect to t we have,

$$\dot{x}\mathbf{i} + \dot{y}\mathbf{j} = -gt\mathbf{j} + \mathbf{c}, \quad \text{where } \mathbf{c} \text{ is a constant vector.} \qquad \textbf{(M1)}$$

We are given that $\dot{x} = u$ and $\dot{y} = v$ when $t = 0 \Rightarrow \mathbf{c} = u\mathbf{i} + v\mathbf{j}$

and
$$\dot{x}\mathbf{i} + \dot{y}\mathbf{j} = u\mathbf{i} + (v - gt)\mathbf{j}. \qquad \text{(i)} \qquad \textbf{(A1 cao)}$$

Integrating again with respect to t, we have

$$x\mathbf{i} + y\mathbf{j} = ut\mathbf{i} + (vt - \tfrac{1}{2}gt^2)\mathbf{j} + \mathbf{b},$$

$$\textbf{(M1)}$$

where $\mathbf{b}$ is another constant vector.

Since P is at O when $t = 0$, we have $x = 0$ and $y = 0$ when $t = 0 \Rightarrow \mathbf{b} = \mathbf{0}$

$$\Rightarrow \quad x\mathbf{i} + y\mathbf{j} = ut\mathbf{i} + (vt - \tfrac{1}{2}gt^2)\mathbf{j}. \qquad \text{(ii)} \qquad \textbf{(A1 cao)}$$

At both O and A, $y = 0$ and (ii) $\Rightarrow vt - \tfrac{1}{2}gt^2 = 0$. This implies that $t = 0$ (at O) and $t = 2v/g$ (at A). The time taken by P to move from O to A, the time of flight, is $2v/g$. $\qquad$ **(A1 cao)**

Since (ii) $\Rightarrow x = ut$, it follows that the distance OA, the horizontal range of P, is $2uv/g$. $\qquad$ **(A1 cao)**

At H, $\dot{y} = 0$ and it follows from (i) that when P is at H, $t = v/g$, half the time required for P to move from O to A.

Also at H, (ii) $\Rightarrow y = v^2/(2g) = HG$, that is the greatest height reached by P above the level of OA is $v^2/(2g)$. $\qquad$ **(A1 cao)**

Note that we have solved this problem from first principles, i.e. without quoting formulae. This is usually to be recommended.

---

2 (a) Using the data as shown in the diagram of question 1, we know that $y = 0$ at both O and A. That is, the values of x given by $x - x^2/40 = 0$ correspond to O and A, and these are $x = 0$ and $x = 40$. The distance OA is 40 m. $\qquad$ **(M1 A1)**

(b) Since $y = x - x^2/40$, we have $\dfrac{dy}{dx} = 1 - x/20$. Since at O, $x = 0$, the gradient of the path of P at O is therefore 1 and the angle of projection is $45°$. This makes the horizontal and vertical components of the initial velocity of P equal, to U m s^{-1}, say. $\qquad$ **(M2)**

We now express the time taken to reach A from O in two different ways by considering the motion of P horizontally and vertically. $\qquad$ **(M1)**
Horizontally, time to reach A from $O = 40/U$ s. $\qquad$ **(A1)**
Vertically, time to reach A from $O = 2U/g$ s. $\qquad$ **(A1)**
Since these are equal, $40/U = 2U/10$ and $U = 10\sqrt{2}$. $\qquad$ **(A1 cao)**

---

3 If $\tan \alpha = \tfrac{5}{12}$, $\cos \alpha = \tfrac{12}{13}$ and $\sin \alpha = \tfrac{5}{13}$. $\qquad$ **(B1)**

(a) The vertical component of the initial velocity of P is $65 \sin \alpha$ m s^{-1} =

25 m s^{-1}. Taking the coordinate axes Ox, Oy as shown in the diagram the y-coordinate of C is -70.

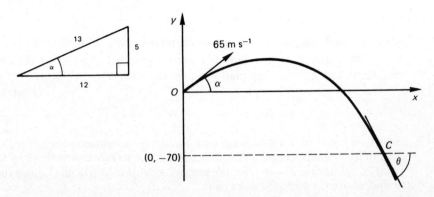

Using $s = ut + \frac{1}{2}at^2$ for the vertical motion of P from O to C, we have $s = -70, u = 25$ and $a = -10$ and this gives the quadratic equation in t:

$$-70 = 25t - 5t^2 \qquad \textbf{(M1 A1)}$$

$$\Leftrightarrow \quad t^2 - 5t - 14 = 0 \quad \Leftrightarrow \quad (t-7)(t+2) = 0 \quad \Rightarrow \quad t = 7$$

($t = -2$ does not apply).
The time taken by P to reach C from O is 7 s. **(M1 A1 cao)**

(b) The horizontal component of the initial velocity of P is $65 \cos \alpha \text{ m s}^{-1} = 60 \text{ m s}^{-1}$. In 7 s, P will move from O to C and cover a horizontal distance $7 \times 60 \text{ m} = 420 \text{ m}$.

(M1 A1 FT)

(c) Consider the horizontal and vertical components of the velocity of P at C. The horizontal component is 60 m s^{-1}, unchanged throughout the motion. **(B1)**
The vertical component is $(25 - 7 \times 10) \text{ m s}^{-1} = -45 \text{ m s}^{-1}$, the negative sign indicates that this component is downwards. **(B1)**
Magnitude of the velocity of P is C is $(60^2 + 45^2)^{1/2} \text{ m s}^{-1} = 75 \text{ m s}^{-1}$. **(M1)**
Direction of velocity of P at C is $\tan^{-1}(45/60) = \tan^{-1}(3/4)$ to the horizontal.

(A1 cao)

(See the diagram where this angle is marked as θ.)

Examples 3 Simple harmonic motion

1 A particle P of mass 0.1 kg moves along Ox with simple harmonic motion of period 0.25 s. The distance between the two points at which it is at

instantaneous rest is 0·24 m. Find the greatest magnitude of (a) the velocity of P, (b) the acceleration of P, (c) the force acting on P.

(8 marks)

2 A load of mass m is placed on a horizontal platform which performs a vertical simple harmonic motion of period T and amplitude a. Show that the load remains in contact with the platform provided that $T \geqslant 2\pi\sqrt{(a/g)}$.

(8 marks)

3 A particle P moves in a straight line with a simple harmonic motion of period T, amplitude a and centre O. Find the time for P to move directly from the point H, distance $a/3$ from O, to the point K, distance $4a/5$ from O, given that H and K are on opposite sides of O.

(6 marks)

4 On a certain day, low water for a harbour occurs at 1130 hours and high water occurs at 1750 hours, the corresponding depths being 4 m and 11 m. Assuming that the tidal rise and fall of the water level is simple harmonic, find the earliest time, to the nearest minute, that a ship drawing 8 m can enter the harbour.

(9 marks)

Hints and solutions for Examples 3

1 Without loss of generality we take O to be the centre of the motion. The period is

$$\frac{2\pi}{n} = 0·25 \text{ s} \quad \Rightarrow \quad n = 8\pi \text{ s}^{-1}. \qquad \textbf{(B1)}$$

The amplitude $a = \frac{1}{2} \times 0·24 \text{ m} = 0·12 \text{ m}$. **(B1)**

(a) The greatest speed of P is na when P is at O.

$$\text{Greatest speed} = 0·12 \times 8\pi \text{ m s}^{-1} \approx 3·02 \text{ m s}^{-1}. \quad \textbf{(M1 A1)}$$

(b) The acceleration has its greatest value $n^2 a$ at an extremity of the motion, when $x = a$,

$$\Rightarrow \quad |\text{greatest acceleration}| = 64\pi^2 \times 0·12 \text{ m s}^{-2}$$

$$\approx 75·8 \text{ m s}^{-2}. \qquad \textbf{(M1 A1)}$$

(c) The greatest magnitude of the force acting on P occurs when P has its greatest acceleration and is therefore

$$0·1 \times 64\pi^2 \times 0·12 \text{ N} \approx 7·58 \text{ N}. \qquad \textbf{(M1 A1)}$$

2

$$2\pi/n = T \quad \Leftrightarrow \quad n = 2\pi/T. \tag{M1}$$

$$x = a \cos nt. \tag{B1}$$

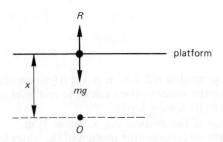

Motion of the load $\updownarrow$,

$$R - mg = m\ddot{x} \tag{M1 A1}$$

$$\Rightarrow \quad R = m(g - n^2 a \cos nt) \tag{M1}$$

$$\Rightarrow \quad R_{min} = m(g - n^2 a). \tag{A1}$$

Load remains in contact with platform if $R_{min} \geqslant 0$, that is if

$$g \geqslant n^2 a \quad \Rightarrow \quad n \leqslant \sqrt{(g/a)} \tag{M1}$$

$$\Rightarrow \quad T \geqslant 2\pi\sqrt{(a/g)}. \tag{A1 cao}$$

Explanation

The first two lines give (i) the relation between the n of our simple harmonic motion formula and the given time of oscillation T, and (ii) a formula for the upward displacement x of the platform from its mean position at time t.

The vertically upward equation of motion of the particle gives R, the magnitude of the force *exerted on the load by the platform* at time t. For the load to remain on the platform R must not be negative. (If R became negative, the platform, on a downward part of its motion, would leave the load behind.)

---

3 Consider the motion as the projection of the motion of a particle moving round the circle $x^2 + y^2 = a^2$ with uniform angular speed ω. **(B1)**

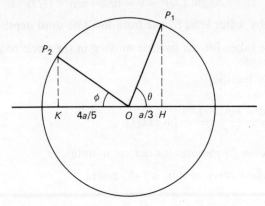

The period $T = 2\pi/\omega$ and so $\omega = 2\pi/T$. **(M1)**

The time from P_1 to P_2, and therefore from H to K, is $(\angle P_1OP_2)/\omega$, **(M2)**

$$\Rightarrow \quad \text{required time} = (\pi - \theta - \phi)/\omega$$

$$= T[\pi - \cos^{-1}\tfrac{1}{3} - \cos^{-1}\tfrac{4}{5}]/(2\pi) \qquad \textbf{(A1)}$$

$$\approx 0.20T. \qquad \textbf{(A1 cao)}$$

---

4 The period of the motion is $2 \times 6\frac{1}{3}$ h $= 12\frac{2}{3}$ h (remembering that the period is the *total* time for the water to rise from its lowest level to its highest level and then fall again to its lowest level). **(B1)**

The amplitude of the motion is $\frac{1}{2} \times 7$ m $= 3\frac{1}{2}$ m. **(B1)**

In the diagram we consider the motion of the water level in the harbour as the projection of the motion of a particle moving with constant angular speed in a circle of radius $3\frac{1}{2}$ m and completing each revolution in time $12\frac{2}{3}$ hours. The points L and P correspond to the depths of water at low tide and when the ship can first enter the harbour. **(M2)**

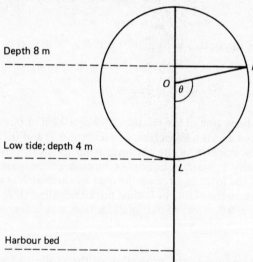

$$\text{Angle } LOP = \theta = \pi/2 + \sin^{-1}(1/7). \qquad \textbf{(A1)}$$

Time taken by water level to rise from low tide until depth is 8 m

$=$ time taken for the particle moving in the circle to go from L to P

$$= \frac{\theta}{2\pi} \times \text{(period)}$$

$$= \left[\frac{\pi/2 + \sin^{-1}(1/7)}{2\pi}\right] \times (12\tfrac{2}{3}) \text{ h} \qquad \textbf{(M1 A2, 1 or 0)}$$

$$\approx 3 \text{ hours } 27 \text{ minutes (to nearest minute).}$$

Time for earliest entry of ship is 1457 hours. **(A1 cao)**

9.3 Short-question tests

Short-question test 5

Time allowed: 40 minutes

1   A particle P is placed on a rough plane inclined at an angle θ to the horizontal. When a force of magnitude Q acts on P upwards along the line of greatest slope of the plane through P, the particle is on the point of moving down the plane. The coefficient of friction between P and the plane is μ. When the force acting on P is increased to magnitude $2Q$ and acts in the same direction, P is on the point of moving up the plane. Prove that $3\mu = \tan \theta$.

(5 marks)

---

2   A caravan, of mass 600 kg, is connected to a car, of mass 900 kg, by means of a light inextensible horizontal tow-bar. The resistive forces to the motion of the caravan and the car are 150 N and 300 N respectively. The caravan and car move along a straight horizontal road with acceleration of magnitude 0.4 m s^{-2}. Calculate, in N,

(a) the magnitude of the tractive force exerted by the engine of the car,

(b) the magnitude of the tension in the tow-bar.

(9 marks)

---

3   At speed v, the resistance to the motion of a particle, of mass m, moving in a straight line, is of magnitude mkv, where k is a constant. Given that the particle starts with speed u from a point O, find the time taken and the distance covered before the particle is moving with speed $\frac{1}{2}u$.

(11 marks)

---

4   A particle is projected horizontally with speed $\sqrt{(8gh)}$ from a point A, situated at a height h above horizontal ground. Given that the particle moves freely under gravity, find, in terms of h, the horizontal distance covered by the particle before it first strikes the ground.

(6 marks)

Hints and solutions for short-question test 5
In each question draw a diagram.

1   When P is on the point of moving down the plane (friction is limiting and acts up the plane)

$$Q = mg \sin \theta - \mu mg \cos \theta. \qquad \textbf{(M1 A1)}$$

When P is on the point of moving up the plane (friction is limiting and acts down the plane)

$$2Q = mg \sin \theta + \mu mg \cos \theta. \qquad \textbf{(M1 A1)}$$

Eliminate Q from these equations and the result follows. **(A1 cao)**

---

2 (a) Considering car and caravan as one body, apply Newton's Second Law.

$$\text{Tractive force} = \big[(600 + 900)0.4 + (300 + 150)\big] \text{ N} \quad \textbf{(M2 A2)}$$
$$= 600 \,\text{N} + 450 \text{ N} = 1050 \text{ N}. \qquad \textbf{(A1 cao)}$$

(b) Considering caravan alone, apply Newton's Second Law.

$$\text{Tension in tow-bar} = (600 \times 0.4 + 150) \text{ N} \qquad \textbf{(M1 A2)}$$
$$= 390 \text{ N}. \qquad \textbf{(A1 cao)}$$

---

3 We have
$$m \frac{dv}{dt} = -mkv, \quad \Rightarrow \quad \int \frac{1}{v}\,dv = \int -k\,dt. \quad \textbf{(M1 A1), (M1)}$$

So
$$\ln v = C - kt. \qquad \textbf{(A1)}$$

When $v = u, t = 0$

$$\Rightarrow \quad C = \ln u \quad \text{and} \quad kt = \ln u - \ln v. \qquad \textbf{(M1)}$$

When $v = u/2$,

$$t = \frac{1}{k}\big[\ln u - \ln (u/2)\big] = \frac{1}{k}\ln 2. \qquad \textbf{(A1 cao)}$$

Particle is moving with speed $u/2$ after time $\dfrac{1}{k}\ln 2$.

Also
$$mv \frac{dv}{dx} = -mkv, \quad \Rightarrow \quad \int 1\,dv = -k\int 1\,dx. \qquad \textbf{(M1), (M1)}$$

So
$$v = C' - kx. \qquad \textbf{(A1)}$$

When $v = u, x = 0$

$$\Rightarrow \quad C' = u \quad \text{and} \quad kx = u - v. \qquad \textbf{(M1)}$$

When $v = u/2$,

$$x = \frac{1}{k}(u - u/2) = \frac{u}{2k}. \qquad \textbf{(A1 cao)}$$

Particle moves through distance $u/(2k)$ in reducing speed from u to $u/2$.

---

4 The initial vertical component of velocity is zero.   **(B1)**
Particle falls a distance h (from rest) in time $\sqrt{(2h/g)}$.   **(M1 A1)**
The horizontal component of the velocity of the particle is constant and equal
to $\sqrt{(8gh)}$.   **(B1)**
Horizontal distance travelled before striking the ground

$$= \sqrt{(8gh)} \times \sqrt{(2h/g)} = 4h. \qquad \textbf{(M1 A1 cao)}$$

Short-question test 6

Time allowed: 35 minutes

1 A particle P is moving with simple harmonic motion in a straight line between
the extreme points A and B, and the mid-point of AB is O. The greatest
acceleration of P is f and the greatest speed of P is u. Find, in terms of u and f,

(a) the distance OA,

(b) the time taken for P to move directly from A to B.

(5 marks)

---

2 A light elastic string AB, obeying Hooke's Law, of natural length a and
modulus of elasticity $4mg$, has the end A tied at a fixed point and a particle, of
mass m, is attached to B. The particle is released from rest when placed at the
side of A and falls vertically. Find the greatest displacement of B from A in the
ensuing motion of the particle.

(7 marks)

---

3 A uniform right circular cone has base radius $2a$ and height $12a$. The cone is
placed with its base on a rough plane inclined at an angle θ to the vertical.
Given that the cone does not slip but that equilibrium is on the point of
breaking down owing to the cone toppling over, find the numerical value of
$\tan \theta$.

(4 marks)

---

4 An unbiased cubical die is to be thrown twice. The score X is found by taking
the difference between the larger number and the smaller number obtained on
the two throws. If the number obtained on the two throws is the same, then
the score is recorded as 0. Calculate the probabilities that $X = 0, 1, \ldots, 5$.
Hence find the probability that X will be less than 2.

(8 marks)

Hints and solutions for short-question test 6

1 The amplitude of the motion is a, where $OA = OB = a$.

We have $\qquad\qquad f = \omega^2 a,\qquad$ and $\qquad u = \omega a.$ **(B1), (B1)**

(a) Solving these equations, $a = u^2/f.$ **(A1 cao)**
(b) Time from A to B = half the period = $\pi/\omega = \pi u/f.$ **(M1 A1 cao)**

---

2 Using the conservation of energy, we observe that the greatest extension occurs when the gain in elastic energy is the same as the loss in potential energy. **(B1)**

Let us suppose extension of string is then x.

$$\text{Energy}\ \Rightarrow\ \tfrac{1}{2}.\,4mgx^2/a = mg(a + x)\quad\textbf{(M1 A2, 1 or 0)}$$
$$\Rightarrow\ 2x^2 - ax - a^2 = 0$$
$$\Rightarrow\ (2x + a)(x - a) = 0\ \Rightarrow\ x = a.\ \textbf{(M1 A1)}$$

Greatest displacement of B from $A = a + a = 2a$. **(A1 cao)**

---

3 The centre of mass G of the cone is at a distance $3a$ from centre of base. **(B1)**
Critical position for toppling is shown in the diagram (vertical broken line) **(M2)**

$\Rightarrow\ \tan\theta = 2a/(3a) = \tfrac{2}{3}.$ **(A1 cao)**

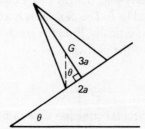

---

4 Work out the following table for the 36 simple events in the possibility space and their associated probabilities.

X	0	1	2	3	4	5
$P(X)$	6/36	10/36	8/36	6/36	4/36	2/36

(Table M2 A3, 2, 1 or 0)

$$P(X < 2) = P(0) + P(1)\qquad\textbf{(M1 A1)}$$
$$= \tfrac{6}{36} + \tfrac{10}{36} = \tfrac{4}{9}.\qquad\textbf{(A1 cao)}$$

---

10 Probability and statistics

10.1 Introduction

If you are doing a full statistics course, at least some of your time will be spent on projects, since statistics is an essentially practical subject. The kinds of questions set in examination papers will be on the theoretical side of the subject, particularly on the use of probability distributions. Sometimes, however, you are asked to give brief explanations of particular points. Be sure to keep such explanations *brief*. Marks are usually awarded on the basis of one mark for each important point made, and usually one or two sentences are enough for this. You will get no more marks for half a page of 'waffle'.

To take an example, you could get the question: 'State the conditions under which the Poisson distribution may be used as an approximation to the binomial distribution. Give an example of the use of this approximation.'

A suitable answer would be as follows: 'The Poisson distribution may be used as an approximation to the Binomial distribution $B(n, p)$ provided that n is large and p is small (or n is large and $np < 5$).' One mark will have been allowed for each of these points so this sentence will gain you two marks.

The example you give should be from your project work. Describe it briefly making sure you bring out the conditions already mentioned above. It should be possible to do this in two or three sentences.

As part of your revision it is worth while noting down the principal features of each of the probability functions you have studied so that you are ready for a question such as the above. You will find the book *Statistics* in this series helpful in this connection.

10.2 Short-question tests

Short-question test 7   (statistics and probability)

Time allowed: 45 minutes

1   A continuous random variable X has probability density function f where

$$f(x) = kx(1 - x) \qquad \text{for } 0 \leqslant x \leqslant 1,$$

$$f(x) = 0, \qquad\qquad \text{otherwise.}$$

Find the value of k and sketch the graph of the function.

(7 marks)

2 The equations of the two regression lines of a bivariate distribution are

$$y = 1 \cdot 2x + 0 \cdot 8,$$

$$x = 0 \cdot 4y + 0 \cdot 2.$$

Calculate the arithmetic mean of each distribution and the product–moment correlation coefficient between the two distributions.

(7 marks)

---

3 A machine is designed to produce screws of length 2 cm. A random sample of 900 screws produced by the machine is found to have a mean length of 2·01 cm and standard deviation 0·06 cm. Estimate the standard error of the mean and obtain an approximate 95 per cent confidence interval for the mean of the whole output of this machine.

(8 marks)

---

4 A manufacturer produces nylon stockings and there is a probability that one pair in every 200 will be defective. A shop orders 600 pairs of stockings from this manufacturer. Estimate the probability that they will get more than 3 defective pairs.

(8 marks)

Solutions and mark scheme for short-question test 7

1
$$\int_0^1 kx(1 - x)\, \mathrm{d}x = k(\tfrac{1}{2} - \tfrac{1}{3}) = k/6.$$

The total probability must be one and so

$$k/6 = 1$$

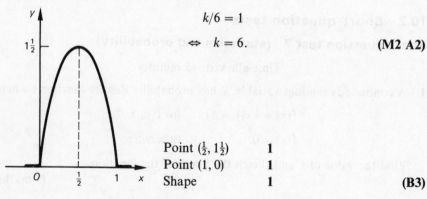

$$\Leftrightarrow k = 6. \qquad \textbf{(M2 A2)}$$

Point $(\tfrac{1}{2}, 1\tfrac{1}{2})$	1
Point $(1, 0)$	1
Shape	1

(B3)

2

$$\bar{y} = 1{\cdot}2\bar{x} + 0{\cdot}8,$$
$$\bar{x} = 0{\cdot}4\bar{y} + 0{\cdot}2$$
$$\Rightarrow \quad \bar{x} = 1, \quad \bar{y} = 2. \qquad \textbf{(M1 A2)}$$

Product moment coefficient

$$r^2 = \frac{s_{xy}^2}{s_x^2 s_y^2} = 1{\cdot}2 \times 0{\cdot}4 = 0{\cdot}48.$$

$$r = 0{\cdot}69. \qquad \textbf{(M2 A2 cao)}$$

---

3 n large, $\sigma = 0{\cdot}06$. $\qquad$ **(B1)**

Distribution of $\bar{x}$ approximately normal. $\qquad$ **(B1)**

Standard error $= \dfrac{0{\cdot}06}{\sqrt{900}} = \dfrac{0{\cdot}06}{30} = 0{\cdot}002$. $\qquad$ **(M1 A1)**

Interval for 95 per cent confidence is

$$2{\cdot}01 \pm 0{\cdot}002 \times 1{\cdot}96 = 2{\cdot}01 \pm 0{\cdot}003\,92$$
$$= 2{\cdot}006\,08 \text{ to } 2{\cdot}013\,92. \qquad \textbf{(M2 A2)}$$

---

4 1 in 200 defective $\Rightarrow p = 0{\cdot}005$. $\qquad$ **(B1)**

n large and p small, Poisson distribution. $\qquad$ **(B1)**

$np = 600 \times 0{\cdot}005 = 3$. $\qquad$ **(B1)**

$$\text{P}(0, 1, 2, 3 \text{ defective pairs}) = e^{-3} \left(1 + \frac{3}{1!} + \frac{3^2}{2!} + \frac{3^3}{3!} \right)$$
$$= e^{-3} (1 + 3 + 4{\cdot}5 + 4{\cdot}5)$$
$$\approx 0{\cdot}647 \qquad \textbf{(M2 A1)}$$

(or use cumulative probability table)

$$\text{P}(x > 3) = 1 - 0{\cdot}647$$
$$= 0{\cdot}353. \qquad \textbf{(M1 A1 cao)}$$

Short-question test 8   (probability)

Time allowed: 40 minutes

1   The probability that a certain seed will germinate is $\frac{7}{10}$. Five such seeds are sown in a tray. Giving your answer to three decimal places, find the probability that at least 3 of the 5 seeds will germinate.

(6 marks)

---

2   Given that Y is a normal variable, estimate the mean and the variance of Y, if $P(Y > 10) = 0.1$ and $P(9 < Y < 10) = 0.2$.

(6 marks)

---

3   A continuous random variable X has probability density function f, given by

$$f(x) = ax^2 + bx, \qquad \text{for } 0 \leqslant x \leqslant 1,$$
$$= 0, \qquad\qquad \text{otherwise},$$

where a and b are constants.
Given that $E(X) = \frac{1}{2}$, find the values of a and b.

(6 marks)

---

4   The number of deaths due to a certain rare illness has been found to follow a Poisson distribution with a mean of 0.8 deaths per week. Estimate, to two significant figures, the probability that there will be

(a) at least 2 deaths in a particular week,

(b) exactly 3 deaths in a particular 3-week period due to this illness.

(7 marks)

10.3 Long-question tests

Paper 11 (probability and statistics)

Time allowed: $2\frac{1}{2}$ hours

Full marks may be obtained for answers to SIX questions.
All questions carry equal marks.

1 (a) Two events A and B are mutually exclusive. Given that $P(A) = \frac{1}{5}$ and $P(A \cup B) = \frac{1}{2}$, find $P(B)$.

  (b) Two events C and D are independent. Given that $P(C) = \frac{1}{4}$ and $P(C \cup D) = \frac{1}{3}$, find $P(D)$.

  (c) A circular disc, of radius r, is thrown at random onto a large board which is divided into squares, each of side $2a$, where $a > r$. Find the probability that the disc comes to rest entirely within one square.

---

2 Explain briefly the circumstances under which a normal distribution may be used as an approximation to a binomial distribution.

  In a factory making children's balls, two machines produce red and green balls respectively in the ratio 3:2. The balls roll into a common container in which they are thoroughly mixed and from which they are packaged into bags of ten, the balls for each bag being randomly chosen. Find the mean and variance of the number per bag of green balls.

  Find the probabilities, to three decimal places, that the number of green balls in a particular bag will be

  (a) one or more,   (b) exactly three.

  Find also an approximate value for the probability that, in a total of 40 bags, there will be more than 170 green balls.

---

3 Describe briefly, with the aid of a diagram, the characteristics of the normal distribution.

  A machine produces metal pins whose lengths are normally distributed with mean 1 cm and standard deviation 0·01 cm. The pin is only suitable for its purpose if its length lies between 0·994 cm and 1·016 cm.

  (a) Find the probability that a randomly chosen pin is satisfactory. Find also the expected number of satisfactory pins in a random sample of 1000 pins.

  (b) Given that the machine is capable of sorting out those pins which are of length greater than 1 cm, find the probability that a randomly chosen pin from this selection is satisfactory.

---

4 A random variable X has geometric distribution with probability parameter p,

i.e. $$P(X = r) = p(1 - p)^{r-1}, \qquad r = 1, 2, 3, \dots$$

Write down the mean and the variance of the distribution.

A board game is such that a player must throw a six with a single fair die in order to start playing. Let X denote the number of throws a particular player makes in order to start.

(a) Write down the probability function of X.

(b) Find the mean and standard deviation of the number of turns required to start.

(c) Find the probability that the player will need at least five turns to start.

(d) Find the smallest value of n for which there is a probability of at least 0·8 that the player will need only n or fewer turns to start.

---

5 A school has two separate buildings, each with its own telephone switchboard. During the period 1 p.m. to 2 p.m. the numbers of incoming calls on each board, A and B, are independent and have Poisson distributions with parameters 6 and 4 respectively. Find, to two significant figures, the probability that during this period, on a randomly chosen day, there will be

(a) exactly 6 calls to switchboard A,
(b) at least 4 calls to switchboard B,
(c) exactly 3 calls to the school altogether.

---

6 The continuous random variable X has the exponential distribution whose probability density function is given by

$$f(x) = \lambda e^{-\lambda x}, \qquad x \geqslant 0,$$

$$f(x) = 0, \qquad \text{otherwise},$$

where λ is a positive constant. Obtain expressions, in terms of λ, for

(a) the mean $E(X)$ of the distribution,
(b) $F(x)$, the cumulative distribution function.

Car exhausts fitted by a certain company are such that their time of useful life in months, X, has the above exponential distribution with $\lambda = 0·01$. Find, to three significant figures, the probability that an exhaust fitted by this company will last at least two years.

The company offers a guarantee of free replacement for exhausts which last less than m months. Assuming that the company would not wish to replace more than one in ten exhausts without charge, find m.

---

7   Explain briefly what is meant by a 98 per cent confidence interval for a population mean. An experiment is being conducted to estimate the length of the circumference of a circular disc. The disc is rolled along a straight line and the length recorded 10 times. The lengths found are known to be normally distributed with a mean equal to the true length and standard deviation equal to 0·3 cm.

(a) Given that the mean of the 10 results was 12·2 cm, determine a 98 per cent confidence interval for the true length.

(b) Determine the smallest number of independent measurings which should be made in order that the 98 per cent confidence interval for the true length of the circumference has a width of less than 0·2 cm.

---

8   (a) The table shows the rankings given to 10 singers by two adjudicators in a music competition.

	A	B	C	D	E	F	G	H	J	K
Rank given by 1st judge	1	2	3	4	5	6	7	8	9	10
Rank given by 2nd judge	2	6	4	3	1	7	5	10	8	9

Calculate Spearman's coefficient of rank correlation between the two judges. Using a table of critical values, comment on the significance of your result, stating clearly the null hypothesis used.

(b) The table shows the marks of 6 candidates in a mathematics examination in which there are two papers, I and II.

Paper I	33	41	43	53	60	70
Paper II	28	36	51	40	58	72

Using x and y to represent marks for Papers I and II respectively, obtain the equation of the regression line of y on x.

A seventh candidate gained 46 on Paper I, but was absent for Paper II. Estimate what mark he would probably have got for Paper II.

Solutions and mark schemes for Paper 11

1 (a) $\quad$ P(A) + P(B) = P$(A \cup B)$ $\quad$ as A and B mutually exclusive. **(M2)**

$$\tfrac{1}{5} + P(B) = \tfrac{1}{2}$$

$$\Rightarrow \quad P(B) = \tfrac{3}{10}. \tag{A1}$$

(b) $\quad$ P(C) . P(D) = P$(C \cap D)$ $\quad$ as C and D independent. **(M2)**

$$P(C \cup D) = P(C) + P(D) - P(C \cap D) \tag{M2}$$

$$\Rightarrow \quad P(C \cup D) = P(C) + P(D) - P(C) . P(D) \tag{M1}$$

$$\Rightarrow \quad \tfrac{1}{3} = \tfrac{1}{4} + P(D) - \tfrac{1}{4}P(D) \tag{A1}$$

$$\Rightarrow \quad P(D) = \tfrac{1}{9}. \tag{A1 cao}$$

(c) If the disc comes to rest entirely within a square S, then the centre O of the disc must lie within the square S_1 which has the same centre as S and is of side $2a - 2r$. But O is equally likely to come to rest at any point on the board. The required probability is therefore $\qquad$ **(B3)**

$$\frac{\text{Area of } S_1}{\text{Area of } S} = \frac{(2a - 2r)^2}{4a^2} = \left(1 - \frac{r}{a}\right)^2. \tag{M2 A2}$$

---

2 n must be large and p not near 0 or 1. $\qquad$ **(B2)**

Mean = np = 4, variance = npq = $4 \times 0.6 = 2.4$. $\qquad$ **(A1 A2)**

(a) $\qquad$ P$(r \geqslant 1)$ = $1 - $ P(0) = $1 - (\tfrac{3}{5})^{10} \approx 0.994.$ $\qquad$ **(M2 A1)**

(b) $\qquad$ P$(r = 3)$ = $^{10}C_3 \, (\tfrac{2}{5})^3(\tfrac{3}{5})^7 = \tfrac{64}{3}(0.6)^9 \approx 0.215.$ $\qquad$ **(M1 A1)**

With 40 bags n is large and p is suitable, so distribution is approximately normal, N(160, 96). $\qquad$ **(M1 A2)**

$$P(X > 170) = 1 - \Phi\left(\frac{170.5 - 160}{\sqrt{96}}\right), \quad \text{making the 0.5} \tag{M2}$$
$$\text{continuity correction,}$$

$$\approx 1 - 0.858 \tag{M1}$$

$$= 0.142. \tag{A1 cao}$$

---

3 Characteristics are symmetry, $f(x) \to 0$ as $n \to \pm \infty$, diagram should show typical bell shape. $\qquad$ **(B3)**

Let X denote the length of a randomly chosen pin.

Then distribution of X normal, $N(1, 0.01^2)$. **(M1)**

$$Z = \frac{X - 1}{0.01} \quad \text{for } N(0, 1). \quad \textbf{(M1)}$$

(a) $\qquad$ P(pin satisfactory) $= P(0.994 < X < 1.016)$

$\qquad\qquad\qquad\qquad = P(-0.6 < Z < 1.6)$ $\qquad$ **(M2 A2)**

$\qquad\qquad\qquad\qquad = \Phi(1.6) - \Phi(-0.6)$

$\qquad\qquad\qquad\qquad = 0.671 \quad$ (to three decimal places). $\quad$ **(A2)**

Expected number of satisfactory pins $= 671$. $\qquad$ **(A1)**

(b) Probability of a satisfactory pin which is longer than 1 cm is

$$P(1 < X < 1.016) = \Phi(1.6) - \Phi(0) \qquad \textbf{(M2)}$$

$$\approx 0.9452 - 0.5$$

$$= 0.4452. \qquad \textbf{(A1)}$$

Probability of length of pin longer than 1 cm, i.e. $P(X > 1)$, $= 0.5$.
P(pin satisfactory $| > 1$ cm) $= 0.4452/0.5 = 0.890$ (to three decimal places).

$\qquad\qquad$ **(M1 A1)**

---

4 Mean $= 1/p$, variance $= (1 - p)/p^2$. $\qquad$ **(B1 B2)**

(a) Probability distribution is

$$P(X = r) = (\tfrac{1}{6})(\tfrac{5}{6})^{r-1}, \qquad r = 1, 2, 3, \ldots . \qquad \textbf{(M2)}$$

(b) Mean $= 6$,

$$\text{Variance} = (1 - \tfrac{1}{6})/(\tfrac{1}{36}) = 30 \qquad \textbf{(A1)}$$

$\Rightarrow \quad$ standard deviation $= 5.48 \quad$ (to two decimal places).

$\qquad\qquad$ **(M1 A1)**

(c) P(at least 5 turns needed) $= \displaystyle\sum_{r=5}^{\infty} (\tfrac{1}{6})(\tfrac{5}{6})^{r-1}$ $\qquad$ **(M2)**

$$= (\tfrac{1}{6})(\tfrac{5}{6})^4 [1 + \tfrac{5}{6} + (\tfrac{5}{6})^2 + \ldots]$$

$$= (\tfrac{5}{6})^4 \approx 0.48 \quad \text{(to two decimal places).}$$

$\qquad\qquad$ **(M1 A1)**

(d) P(n or fewer turns needed) = 1 - P[at least (n + 1) turns] **(M2)**

$$\Rightarrow \quad 1 - (\tfrac{5}{6})^n \geqslant 0.8$$

$$\Rightarrow \quad (\tfrac{5}{6})^n \leqslant 0.2$$

$$\Rightarrow \quad n \geqslant (\lg 0.2)/(\lg \tfrac{5}{6})$$

$$\Rightarrow \quad n = 9. \qquad \textbf{(M1 A2)}$$

---

5 (a) Let X denote the number of calls to A between 1 and 2 p.m.

$$P(X = r) = \frac{e^{-6} 6^r}{r!}, \qquad r = 0, 1, 2, \ldots,$$

$$P(X = 6) = \frac{e^{-6} 6^6}{6!}$$

$$= 0.16 \qquad \text{(to two decimal places). } \textbf{(M2 A2)}$$

(b) Let Y denote the number of calls to B between 1 and 2 p.m.

$$P(Y = 4) = \frac{e^{-4} 4^r}{r!}$$

$$P(Y = 0, 1, 2, 3) = 0.433 \quad \text{(from table of cumulative probabilities)}$$
(M2 A1)

$$P(Y > 4) = 1 - 0.433$$

$$= 0.57 \qquad \text{(to two decimal places). } \textbf{(M1 A1)}$$

(c) $\quad$ P(exactly 3) = P[(A0, B3); (A1, B2); (A2, B1); (A3, B0)]

(M2)

$$= e^{-6} \cdot \frac{e^{-4} \cdot 4^3}{3!} + \frac{e^{-6} \cdot 6}{1!} \cdot \frac{e^{-4} \cdot 4^2}{2!}$$

$$+ \frac{e^{-6} \cdot 6^2}{2!} \cdot \frac{e^{-4} \cdot 4}{1!} + \frac{e^{-6} \cdot 6^3}{3!} \cdot e^{-4} \qquad \textbf{(M4)}$$

$$\approx e^{-10} (166.7)$$

$$= 0.0076 \quad \text{(to two significant figures). } \textbf{(A2)}$$

---

6 (a) $\qquad\qquad E(X) = \lambda \int_0^\infty x\, e^{-\lambda x}\, dx = 1/\lambda.$ $\qquad$ **(M1 A2)**

(b) $\qquad\qquad F(x) = \lambda \int_0^x e^{-\lambda x}\, dx = 1 - e^{-\lambda x}.$ $\qquad$ **(M1 A2)**

$P(X \geqslant 24) = 1 - F(24) = e^{-0.24} = 0.787$ (to three significant figures), i.e. probability of exhaust lasting 2 years $= 0.787$. **(M2 A2)**

$$P(\text{exhaust lasts less than } m \text{ months}) = P(X < m) = F(m) \quad \textbf{(M1)}$$

$$\Rightarrow \quad 1 - e^{-0.01m} \leqslant 0.1$$

$$\Rightarrow \quad e^{0.01m} \leqslant 10/9$$

$$\Rightarrow \quad m \leqslant 10.5. \quad \textbf{(M2 A3)}$$

Hence company could only guarantee for 10 months. **(A1 cao)**

---

7 Explanation should make clear that the interval has a 98 per cent probability of including the population mean. **(M2)**

(a) Mean length 12.2 cm, standard deviation 0.3 cm.
    For interval

$$1 - 2\alpha = 0.98$$

$$\Rightarrow \quad \alpha = 0.01. \quad \textbf{(M1)}$$

98 per cent confidence interval is

$$12.2 \pm 2.33(0.3/\sqrt{10}) \approx 12.2 \pm 0.22 \quad \textbf{(M4 A2)}$$

$$= 11.98 \text{ to } 12.42. \quad \textbf{(A2 cao)}$$

(b) Width 0.2.
    For n experiments

$$SE = 0.3/\sqrt{n}. \quad \textbf{(M1)}$$

$$\text{Width} = 2(2.33) \times \frac{0.3}{\sqrt{n}} < 0.2 \quad \textbf{(M2 A1)}$$

$$\Rightarrow \quad \sqrt{n} > \frac{2 \times 2.33 \times 0.3}{0.2}$$

$$\Rightarrow \quad \sqrt{n} > 6.99 \quad \Rightarrow \quad \text{least } n = 49. \quad \textbf{(M1 A1)}$$

At least 49 measurings needed.

---

8 (a) Spearman's coefficient

$$r_s = 1 - \frac{6\sum d^2}{n(n^2 - 1)}, \qquad \sum d^2 = 46. \quad \textbf{(M1 A1)}$$

$$r_s = 1 - 0.2788$$

$$= 0.721. \tag{A2}$$

Null hypothesis H_0: $r_s = 0$. **(B1)**

From table, $n = 10$, $P = 1.5$ per cent approx. **(B1)**

Can reject H_0 at the 5 per cent level. There appears to be direct correlation between the 2 sets of ranks. **(M2)**

(b)
$$\sum x = 300, \qquad\qquad \sum y = 285.$$

$$\bar{x} = 50 \qquad\qquad\qquad \bar{y} = 47.5. \quad \textbf{(B1 B1)}$$

$$\sum (x - \bar{x})^2 = 928, \qquad \sum (x - \bar{x})(y - \bar{y}) = 983, \quad \textbf{(B1 B1)}$$

$$\sum (x - \bar{x})(y - \bar{y})/\sum (x - \bar{x})^2 = 983/928. \quad \textbf{(A1)}$$

Equation is

$$y - 47.5 = \frac{983}{928}(x - 50) \quad \Rightarrow \quad y = 1.06x - 5.46. \quad \textbf{(M2 A1)}$$

When $x = 46$, $y = 43$ to the nearest whole number, which is the mark he might have got on Paper II. **(A1)**

Paper 12 (probability)

Time allowed: $2\frac{1}{2}$ hours

Answer ALL questions in Section I and FOUR questions in Section II.

SECTION I (48 marks)

Answer ALL questions in this section.

1 A girl tosses a coin three times and a boy tosses a coin twice. Find the probability that the boy gets more heads than the girl.

(5 marks)

---

2 For two events A and B it is known that $P(A) = \frac{3}{5}$, $P(B) = \frac{1}{5}$ and $P(A|B) = \frac{1}{10}$. Find the probability that

(a) both A and B occur,

(b) just one of the events A and B occurs.

(6 marks)

---

3 In a certain population the probability of a girl having blue eyes is $\frac{2}{5}$. A random sample of six girls is selected. Find the probability that from this sample

(a) exactly three girls will have blue eyes,

(b) at least three girls will have blue eyes.

(7 marks)

---

4 In a large consignment of oranges the average number of rotten oranges is 4 per cent. Use the Poisson distribution to estimate the probability that out of a sample of 50 such oranges, taken at random, more than 4 rotten oranges are present.

(7 marks)

---

5 A continuous random variable X takes values x such that $0 \leqslant x \leqslant 2$ and its probability density function f is given by

$$f(x) = ax^2 + bx, \qquad \text{for } 0 \leqslant x < 2,$$

$$f(x) = 0 \qquad\qquad \text{otherwise.}$$

Show that $b = \frac{3}{2}$ and find the value of a. Find the variance of X.

(7 marks)

---

6 Using only the numbers $1, 5, 7, 8$ without repetition, calculate

(a) the number of different numbers exceeding 1000 which can be formed,

(b) the number of odd numbers exceeding 700 which can be formed.

(8 marks)

---

7 In a particular examination there were 1000 candidates and the pass mark was 46. Assuming that the marks were approximately normally distributed with mean 40 and that 330 candidates passed the examination, estimate

(a) the standard deviation of the marks,

(b) the number of candidates who scored a mark between 60 and 70 inclusive.

(8 marks)

SECTION II

Answer FOUR questions only in this section.

8 The following information is known about two events X and Y.

$$P(X) = \tfrac{2}{5}, \quad P(Y \mid X) = \tfrac{7}{10} \quad \text{and} \quad P(X' \cap Y) = \tfrac{3}{10}.$$

Find the following probabilities:

(a) $P(X \cap Y)$, (b) $P(Y)$, (c) $P(X \cup Y)$, (d) $P(X \mid Y)$.

(13 marks)

---

9 A bag contains 6 red, 4 black and 5 white balls, identical except for colour. Three balls are drawn from the bag, one at a time, at random and without replacement. Find the probability that

(a) there will be one ball of each colour,

(b) exactly one ball will be red,

(c) at least one ball will be red,

(d) the second ball drawn will be red,

(e) the second ball drawn will be red, given that the first ball drawn is red.

(13 marks)

---

10 Out of a large batch of mass produced articles, it is known that 1 per cent are defective. A random sample of 40 of these articles is taken. Use (a) the binomial distribution, (b) the Poisson distribution, to estimate for this sample the probability that less than 2 defective articles are present.

(13 marks)

---

11 The continuous random variable X is normally distributed with mean 7 and standard deviation 4. Given that $W = 2X - 5$, find the mean and standard deviation of W.

Estimate:

(a) $P(X < 6)$, (b) $P(6 < W < 10)$.

(13 marks)

---

12 State two conditions which must be satisfied by all probability density functions.

The probability density function f of a random variable X takes values x such that

$$f(x) = 0 \qquad \text{for } |x| > 2,$$
$$f(x) = 3k(4 - x^2) \qquad \text{for } |x| \leqslant 2,$$

where k is a positive constant.

Sketch the graph of f and determine the value of k.

Calculate the probability that X takes a value between -1 and $+1$.

(13 marks)

---

13 A discrete random variable X takes values x, where x is a positive integer. The probability that X takes a value x from $\{1, 2, 3, \ldots\}$ is given by

$$P(X = x) = p(x) = g(1 - g)^{x-1}, \qquad \text{where } 0 < g < 1.$$

Prove that the mean of X is $1/g$ and that the variance of X is $(1 - g)/g^2$.

(Note that this discrete probability function is known as a *geometric distribution*.)

A fair die is to be thrown repeatedly until a six appears. Find the probability that a six will first appear on the third throw. State the mean and the variance of the geometric distribution governing this series of trials.

(13 marks)

Answers

Paper 1 (multiple-choice test)

1	B	6	C	11	B	16	E	21	E	26	D
2	D	7	D	12	E	17	C	22	B	27	B
3	D	8	E	13	D	18	A	23	A	28	A
4	A	9	B	14	B	19	E	24	D	29	E
5	A	10	A	15	C	20	C	25	C	30	C

Paper 2 (multiple-choice test)

1	D	6	B	11	E	16	C	21	E	26	B
2	B	7	E	12	D	17	B	22	C	27	D
3	A	8	A	13	B	18	D	23	C	28	D
4	E	9	C	14	E	19	A	24	A	29	A
5	C	10	C	15	B	20	B	25	A	30	D

Paper 3 (multiple-choice test)

1	B	6	D	11	A	16	C	21	B	26	E
2	D	7	A	12	C	17	E	22	A	27	C
3	C	8	A	13	E	18	A	23	D	28	C
4	E	9	E	14	B	19	D	24	E	29	B
5	C	10	E	15	D	20	B	25	A	30	D

Paper 4 (multiple-choice test)

1	B	6	C	11	D	16	E	21	A	26	B
2	D	7	E	12	A	17	A	22	D	27	B
3	D	8	B	13	C	18	B	23	B	28	A
4	C	9	E	14	B	19	B	24	C	29	C
5	A	10	A	15	D	20	C	25	E	30	E

Paper 7

1 $3, -2, \dfrac{\frac{3}{5}}{x-3} + \dfrac{\frac{2}{5}}{x+2}$

2 $\frac{17}{6} < x < \frac{13}{4}$

3 2520, 720, 360

4 $n^2 + 2^{n+1} - n - 2$

5 $-2(x - 1/x)^{-3}(1 + 1/x^2), \frac{1}{2}x + \frac{1}{4}\ln(1 + 2x) + C$

6 $\pi/24, 25\pi/24$

7 2·5, 1·2

9 $(1, \frac{1}{2})$, $(-1, -\frac{1}{2})$; $\frac{1}{2} \ln (1 + a^2)$

10 (a) $\mathbf{i} - 3\mathbf{j} + 5\mathbf{k}$;

    (b) $\mathbf{r} = \mathbf{i} + 4\mathbf{j} - 3\mathbf{k} + \lambda(\mathbf{i} - 3\mathbf{j} + 5\mathbf{k})$;

    (c) $-2\mathbf{i} + \mathbf{j} + \mathbf{k}$;

    (d) $[\mathbf{r} - (\mathbf{i} + 4\mathbf{j} - 3\mathbf{k})] \cdot [-2\mathbf{i} + \mathbf{j} + \mathbf{k}] = 0$;

    (e) $2\sqrt{6}$

11 $\dfrac{1}{1 - x} + \dfrac{x - 1}{4 + x^2}$, $-\ln (1 - x) + \frac{1}{2} \ln (4 + x^2) - \frac{1}{2} \tan^{-1} (x/2)$,

    $\frac{3}{4} + \frac{5}{4}x + \frac{17}{16}x^2 + \frac{15}{16}x^3 + \frac{63}{64}x^4 + \ldots$

12 (a) $0{\cdot}524$, $3{\cdot}144$;   (b) $\dfrac{2}{y} = \dfrac{5}{2} - x + \dfrac{x^2}{2}$

13 $5 \cos (\theta + 36{\cdot}87°)$; $16{\cdot}3°$, $270°$; $121, 1$; $1, \frac{1}{11}$

14 $0{\cdot}98$

Paper 10

1 $\frac{1}{2}ak\sqrt{17}$

2 $u, \frac{1}{4}$

3 $g = a\,\omega^2 \sin \theta$

4 $24/7$ N, $45/28$ J

6 $h = \sqrt{(3)}r$, $60°$

7 $50 \ln (5/2)$ m

8 (a) $\frac{1}{5}$;   (b) $\frac{1}{10}$;   (c) $\frac{1}{3}$

9 $ut(1 + e^{\omega t}) + \dfrac{2u}{\omega}(1 - e^{\omega t})$, $u, \dfrac{u}{\omega}(3 - e)$

10 $l\sqrt{2}$, $\frac{1}{2}\pi\sqrt{(2l/g)}$

11 $135{\cdot}5°$, $300°$; 140 minutes

12 $13P$, $13a/6$, $36Pa$

13 $Mg\sqrt{(3/2)}$ at $\tan^{-1} \sqrt{(3/5)}$ with horiz., $Mg, (5 - 2\sqrt{5})/10$

14 $16{\cdot}5$ kW, $8°$, $25/36$ m s^{-2}

Short-question test 2

2 $-2 < x \leqslant 0, 2 < x \leqslant 4$

5 $(1, 0)$, $(\frac{1}{4}, -\frac{3}{4})$, $x = \pm\frac{1}{2}$, $y = \frac{1}{4}$

Short-question test 3

1 $58{\cdot}4°$

3 $\mathbf{r} = 2\mathbf{i} - \mathbf{j} + 3\mathbf{k} + \lambda(2\mathbf{i} + 4\mathbf{j} - 5\mathbf{k})$, $(1, -3, 5\frac{1}{2})$

4 $\mathbf{r} = \mathbf{k}$, $\mathbf{i} - \mathbf{j} + \mathbf{k}$, $(\mathbf{r} - \mathbf{k}) \cdot (\mathbf{i} - \mathbf{j} + \mathbf{k}) = 0$

Short-question test 4

2 $\dfrac{-33}{4}$

3 $106, -8\frac{1}{12}$

4 $cx^2 + bx + a = 0$, $a^2cx^2 + (b^3 - 3abc)x + ac^2 = 0$

Short-question test 8

1 $0{\cdot}837$

2 $8{\cdot}309$, $1{\cdot}32$

3 $-6, 6$
4 $0 \cdot 19, 0 \cdot 21$

Paper 12

1 $\frac{3}{16}$
2 (a) $\frac{1}{50}$; (b) $\frac{19}{25}$
3 (a) $0 \cdot 28$; (b) $0 \cdot 46$
4 $0 \cdot 053$
5 $-\frac{3}{4}; 0 \cdot 2$
6 (a) 24; (b) 28
7 (a) $12 \cdot 5$; (b) 52
8 (a) $0 \cdot 28$; (b) $0 \cdot 58$; (c) $0 \cdot 7$; (d) $0 \cdot 483$
9 (a) $0 \cdot 264$; (b) $0 \cdot 475$; (c) $0 \cdot 815$; (d) $0 \cdot 4$; (e) $\frac{5}{14}$
10 (a) $0 \cdot 939$; (b) $0 \cdot 938$
11 $9, 8$; (a) $0 \cdot 401$; (b) $0 \cdot 196$
12 $f(x) \geqslant 0, \int_a^b f(x) \, dx = 1, k = \frac{1}{32}, \frac{11}{16}$
13 $0 \cdot 116$; 6, 30

Index

(Note: This index is not exhaustive, but has been constructed to provide rapid access in looking up particular words, expressions and subject areas contained either in the text or in worked examples.)